THE GREEN COMPUTING BOOK

Tackling Energy Efficiency
at Large Scale

T0225544

Chapman & Hall/CRC
Computational Science Series

SERIES EDITOR

Horst Simon
Deputy Director
Lawrence Berkeley National Laboratory
Berkeley, California, U.S.A.

PUBLISHED TITLES

COMBINATORIAL SCIENTIFIC COMPUTING
Edited by Uwe Naumann and Olaf Schenk

CONTEMPORARY HIGH PERFORMANCE COMPUTING: FROM PETASCALE
TOWARD EXASCALE
Edited by Jeffrey S. Vetter

DATA-INTENSIVE SCIENCE
Edited by Terence Critchlow and Kerstin Kleese van Dam

PETASCALE COMPUTING: ALGORITHMS AND APPLICATIONS
Edited by David A. Bader

FUNDAMENTALS OF MULTICORE SOFTWARE DEVELOPMENT
Edited by Victor Pankratius, Ali-Reza Adl-Tabatabai, and Walter Tichy

THE GREEN COMPUTING BOOK: TACKLING ENERGY EFFICIENCY AT LARGE SCALE
Edited by Wu-chun Feng

GRID COMPUTING: TECHNIQUES AND APPLICATIONS
Barry Wilkinson

HIGH PERFORMANCE COMPUTING: PROGRAMMING AND APPLICATIONS
John Levesque with Gene Wagenbreth

HIGH PERFORMANCE VISUALIZATION:
ENABLING EXTREME-SCALE SCIENTIFIC INSIGHT
Edited by E. Wes Bethel, Hank Childs, and Charles Hansen

INTRODUCTION TO COMPUTATIONAL MODELING USING C AND
OPEN-SOURCE TOOLS
José M. Garrido

INTRODUCTION TO CONCURRENCY IN PROGRAMMING LANGUAGES
Matthew J. Sottile, Timothy G. Mattson, and Craig E. Rasmussen

INTRODUCTION TO ELEMENTARY COMPUTATIONAL MODELING: ESSENTIAL
CONCEPTS, PRINCIPLES, AND PROBLEM SOLVING
José M. Garrido

THE GREEN COMPUTING BOOK

Tackling Energy Efficiency at Large Scale

Edited by

Wu-chun Feng

Virginia Polytechnic Institute
and State University
Blacksburg, USA

CRC Press
Taylor & Francis Group
Boca Raton   London   New York

CRC Press is an imprint of the
Taylor & Francis Group, an **informa** business

A CHAPMAN & HALL BOOK

CRC Press
Taylor & Francis Group
6000 Broken Sound Parkway NW, Suite 300
Boca Raton, FL 33487-2742

First issued in paperback 2020

ISBN-13: 978-1-4398-1987-6 (hbk)
ISBN-13: 978-0-367-65915-8 (pbk)

Library of Congress Cataloging-in-Publication Data

Green computing : large-scale energy efficiency / [edited by] Wu-chun Feng.
    pages cm. -- (Chapman & Hall/CRC computational science ; 21)
  Includes bibliographical references and index.
  ISBN 978-1-4398-1987-6 (hardback)
    1. Information technology--Environmental aspects. 2. Information
technology--Energy consumption. 3. Computer systems--Energy conservation. 4. Green
technology. I. Feng, Wu-chun, 1966-, editor.

QA76.9.E58I58 2013
628--dc23   2013036874

**Visit the Taylor & Francis Web site at
http://www.taylorandfrancis.com**

**and the CRC Press Web site at
http://www.crcpress.com**

Table of Contents

Preface

INTRODUCTION

In 1957, the Soviet Union launched the first Earth-orbiting artificial satellite—Sputnik I—into space. This singular event precipitated the birth of the space age and, more specifically, the U.S.–U.S.S.R. space race. If we fast forward 45 years, the year is 2002, and Japan unveils the first supercomputer that obliterates U.S. domination in supercomputing,*creating such a fervor that the event is dubbed Compute-nik and ignites an "arms race in supercomputing" in which the need for speed is paramount above all else and supercomputing (also referred to as high-performance computing or HPC) becomes increasingly mainstream, as illustrated in Figure 0.1 from the U.S. Council of Competitiveness, and as detailed in a report by the President's Information Technology Advisory Committee (PITAC).

However, this singular focus on speed as a performance metric arguably comes at the expense of other performance metrics, such as efficiency, reliability, and availability, to name a few. These "other" metrics are of particular interest to our *Supercomputing in Small Spaces (SSS)* project, which started in 2001, when being cool was "not cool." Why? With the supercomputing community's focus on speed, supercomputing nodes were not only becoming faster but also consuming and dissipating more power. By applying Arrhenius's equation† to microelectronics or, more generally, computer hardware, every 10°C increase in temperature doubles the failure rate of a given system. This equation was supported by our own informal empirical data at the time, specifically a 128-node cluster that resided in a warehouse and failed approximately once per week during the winter months when the temperature inside the warehouse was 21–23°C and approximately twice per week during the summer months when the temperature was 30–32°C. As a consequence, we learned that by keeping the power draw lower, and

---

*The Japanese supercomputer, called Earth Simulator, delivered a sustained performance that was approximately five times faster than the next-fastest supercomputer in the world.

†Arrhenius's equation notes that for many common chemical reactions at room temperature, the reaction rate doubles for every 10°C increase in temperature.

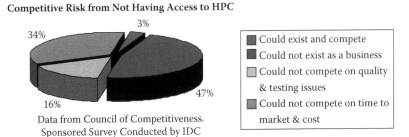

Competitive Risk from Not Having Access to HPC

Could exist and compete
Could not exist as a business
Could not compete on quality & testing issues
Could not compete on time to market & cost

Data from Council of Competitiveness.
Sponsored Survey Conducted by IDC

FIGURE 0.1   The importance of high-performance computing (HPC).

in turn system temperature lower, the efficiency, reliability, and availability of a supercomputer could be significantly improved. Furthermore, with 65% of information technology managers reporting that their websites were unavailable to customers over a 6-month period, and that the cost of the service outages ran as high as $6,450,000/hour for a New York City stockbroker and $1,200,000/hour for Amazon.com, businesses also require HPC (or, more broadly, large- or extreme-scale computing) that delivers near-100% availability with efficient and reliable resource usage in support of e-commerce, enterprise applications, and data centers.

By the late 2000s, the data center community finally realized the need for energy-efficient HPC as the annual energy costs for a data center surpassed annual server purchase costs in 2008, as demonstrated in Figure 0.2. Around this same time frame, the supercomputing community also recognized the importance of both speed and power consumption in the context of tomorrow's exascale computing systems. For example, while past trends indicate that an exascale supercomputer (10^{18} floating-point operations per second or 10^{18} FLOPS) would arrive in 2018, such a system is projected to consume more than 100 megawatts (MW) of power, thus making power consumption the primary design constraint for achieving such exascale performance. Even the floating-point units (FPUs) alone are projected to consume 10 MW of power. Based on these projections, "performance at any cost" is no longer practical.

GREEN SUPERCOMPUTING: PAST, PRESENT, AND FUTURE

While the supercomputing community focused on performance (i.e., speed) in the early 2000s, as exemplified by the Top500 list (http://www.top 500.org/) and the Gordon Bell Awards at the ACM/IEEE (Association for Computing Machinery/Institute of Electrical and Electronics Engineers)

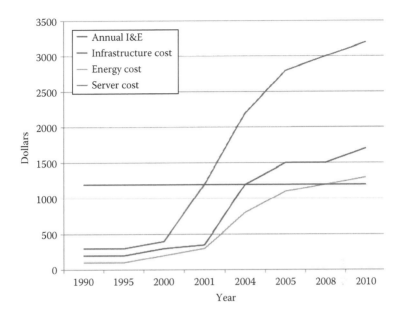

FIGURE 0.2 Annual amortized costs in the data center.

Supercomputing Conference every year, I instead built *MetaBlade*, a 24-node cluster having a footprint of 19×25 in. (or 3.3 ft^2) and consuming a meager 300 watts (W) of power at idle, when booted diskless, in 2001. To put this into perspective, two large-size pizza boxes placed side by side have a footprint of 16×32 in. (or 3.5 ft^2), and the equivalent of three 100 W lightbulbs could power MetaBlade.

How was such a feat possible? Each processor (or central processing unit, CPU) in MetaBlade was a Transmeta TM5800 that ran our customized high-performance code-morphing software, which improved overall performance while maintaining a small thermal power envelope. Specifically, a Transmeta TM5600 CPU by itself dissipated only 6 W under load.

As a follow-up to MetaBlade and as the first major instantiation of the SSS project, MetaBlade was extended to create the *Green Destiny* supercomputer in April 2002 with the tenets of efficiency, reliability, and availability in mind. Green Destiny* was a 240-node cluster with a slightly larger footprint than MetaBlade—24×30 in. (or 5.0 ft^2)—but approximately 14 times

---

*The namesake for Green Destiny (GD) has a multiplicity of origins. First, the destiny of the GD was to be green. Second, GD was based on computer "blade" technology; GD was also the name of the sword blade in *Crouching Tiger, Hidden Dragon*. Third, the names of the groundbreaking supercomputers from the ASCI program at the U.S. Department of Energy were named after different colors; our project chose green.

taller than MetaBlade for a total volume of 32 ft³, which is approximately the size of a small telephone booth. The entire Green Destiny supercomputer consumed only 3.2 kW of power, roughly the equivalent in power to two hairdryers. Due in large part to its very low thermal power envelope, Green Destiny did not have any unscheduled downtime in its 2-year lifetime, despite residing in a dusty 85–90°F warehouse at over 7,000 ft above sea level. Because the CPUs in Green Destiny consumed so little power, the CPUs did not require active cooling, could be closely packed, and provided tremendous savings in terms of operational costs, space constraints, and cooling infrastructure. This audacious feat led to worldwide acclaim, including media coverage in *The New York Times*, CNN, and *PC World*, just to name a few.* However, this feat did *not* lead to the worldwide embracement of the large-scale green computing movement, at least not at the time. It did, however, provide the impetus for significant seminal research in the area of large-scale green computing, as exemplified in this book.

This green book is organized into nine chapters and loosely ordered to start with low-level, hardware-based approaches and traversing up the software stack with increasingly higher-level, software-based approaches. Debuting in November 2004, the IBM Blue Gene approach to green supercomputing sets the stage in Chapter 1, "Low-Power, Massively Parallel, Energy-Efficient Supercomputers," authored by the IBM Blue Gene team. In this chapter, IBM architects focus on low-power, high-performance designs that illustrate how to improve the energy efficiency of a supercomputer by an order of magnitude without any system performance loss in parallelizable applications.

With the export of software-based mechanisms, such as dynamic voltage and frequency scaling (DVFS) and dynamic concurrency throttling (DCT), to control underlying computer hardware, the energy efficiency of a large-scale computing system can be enhanced further *if* these mechanisms are applied appropriately (i.e., at the right time and at the right setting). To that end, Chapter 2, "Compiler-Driven Energy Efficiency," by Kandemir and Srikantaiah seeks to reshape a program either explicitly, by inserting instructions to turn off resources that are not being used, or implicitly, by increasing the idle periods between two usages of the resource. In particular, the authors demonstrate this via compiler-directed energy optimizations in both the I/O (input/output) subsystem and the network on a chip (NoC), which can consume significant power due to on-chip communication.

---

*Green Destiny and an identical replica of it now reside in the Computer History Museum in Mountain View, California, and in the Bradbury Science Museum in Los Alamos, New Mexico, respectively.

As a complement to the compiler-driven approach mentioned, Hsu, Feng, and Poole present their dynamic run-time system that *automatically* schedules the DVFS mechanism at the right voltage and frequency and at the right time to maintain performance while reducing both power and energy consumption. Serendipitously, the work in Chapter 3, "An Adaptive Run-Time System for Improving Energy Efficiency," was in response to criticism of Green Destiny and, in turn, as an outgrowth of the SSS project. Specifically, many criticized Green Destiny for sacrificing too much performance to reduce power consumption, while others complained that Green Destiny used too many pseudoproprietary parts. By leveraging DVFS on commodity processors from AMD, the adaptive run-time system presented in Chapter 3 addresses these criticisms in one fell swoop and has since been adapted to other commodity processors.

Similarly, Chapter 4, "Energy-Efficient Multithreading through Run-Time Adaptation," by Curtis-Maury and Nikolopoulus presents a general framework that predicts the performance impact of dynamically adjusting different power-performance settings and derives predictors to efficiently identify optimal settings at run time. The framework simultaneously leverages both DVFS and DCT across multiple multicore processors. In addition, this chapter provides a survey of methods for power-performance adaptation in run-time systems.

Chapter 5, "Exploring Trade-Offs between Energy Savings and Reliability in Storage Systems," by Butt, Bhattacharjee, Wang, and Gniady not only explores the interactions between energy management and reliability but also studies storage system organization that maximizes energy efficiency and reliability. To support these items, the authors propose a new metric for simultaneously evaluating energy and reliability called the energy-reliability product (ERP) and the basic mechanisms for idle time allocation between energy management and reliability mechanisms to achieve a balance between energy consumption and reliability.

From a more holistic perspective, Chapter 6, "Cross-Layer Power Management," by Wang and Ranganathan addresses the need for coordinated power control across different layers: hardware, firmware, operating system (OS), and application level. The authors identify a key challenge that arises when multiple controllers at different layers interfere with each other due to the lack of adequate coordination between the controllers and then present the benefits of cross-layer power management solutions.

In Chapter 7, "Energy-Efficient Virtualized Systems," Nathuji and Schwan discuss the nexus of two problem domains, energy management and virtualization, which are proving to be critical for cloud computing environments and the data centers that house them. Specifically, the authors articulate how current virtualization systems lack the capability to perform active power management while still supporting quality-of-service requirements. They then describe ways for extending existing virtualization architectures to better support energy efficiency.

As conveyed in its title, "Demand Response for Computing Centers," Chapter 8 by Chase studies the demand response (DR) in computing centers, where DR refers to policies that influence the timing or location of power demand when responding to signals from the electrical supplier about energy production cost or availability. This, in turn, is intended to improve the reliability and efficiency of electrical power grids and the myriad associated "smart grid" initiatives.

Finally, Chapter 9, "Implications of Recent Trends in Performance, Costs, and Energy Use for Servers," by Koomey, Belady, Patterson, Santos, and Lange assesses trends in servers and their impact on data center costs. Specifically, the authors summarize trends in server costs, energy use, and performance and then describe the implications of these trends in the context of the economics of high-density computing facilities.

In addition to the contributing authors of this book, there are many people to thank for their role in its realization. First and foremost, I thank Horst Simon for providing me with the opportunity to create this book as a consequence of my early (and controversial) research in green supercomputing with Green Destiny, its evolution into a software-based approach called *beta* (see Chapter 3) and commercially known as EnergyFit, and the subsequent initiation of the Green500 (http://www.green500.org/). As the Chapman & Hall/CRC Computational Science Series editor, Horst provided guidance in shaping the book and served as a source of inspiration with his pursuit of energy efficiency for large-scale computing systems, as exemplified by his tireless efforts toward realizing the University of California's Computational Research and Theory (CRT) Facility.

Many thanks to Randi Cohen, computer science acquisitions editor for Chapman & Hall/CRC Press, who was critical in the development and production of this book. Without her assistance, support, cajoling, and patience, this book would not have been possible.

Finally, I thank my wife, Annette Feng, and children, Akaela and Kai, for their understanding and patience while I juggled the demands of work, the preparation of this book, and family time.

I hope that this green book inspires you the way that it has inspired the authors and me to write about it and to raise the awareness of *greenness* as a first-order design constraint that is on par with performance. As a tongue-in-cheek comment, we hope that this book will encourage data centers and supercomputing centers to simulate climate change rather than "create it."

Contributors

Christian Belady
Microsoft
Seattle, Washington

Puranjoy Bhattacharjee
Amazon
Seattle, Washington

Ali R. Butt
Virginia Tech
Blacksburg, Virginia

Jeffrey S. Chase
Duke University
Durham, North Carolina

George Chiu
IBM T.J. Watson Research Center
Yorktown Heights, New York

Matthew Curtis-Maury
NetApp
Raleigh, North Carolina

Wu-chun Feng
Virginia Tech
Blacksburg, Virginia

Chris Gniady
University of Arizona
Tucson, Arizon

Chung-Hsing Hsu
Oak Ridge National Laboratory
Oak Ridge, Tennessee

IBM Blue Gene Team®
IBM T. J. Watson Research Center
Yorktown Heights, New York

Mahmut Kandemir
Pennsylvania State University
State College, Pennsylvania

Jonathan G. Koomey
Stanford University
Palo Alto, California

Klaus-Dieter Lange
Hewlett-Packard
Houston, Texas

Ripal Nathuji
NeoTek Labs
Austin, Texas

Dimitris Nikolopoulos
Forth Institute of Computer Science
Crete, Greece

Michael Patterson
Stanford University
Palo Alto, California

Stephen W. Poole
Oak Ridge National Laboratory
Oak Ridge, Tennessee

Parthasarathy Ranganathan
Google
Mountain View, California

Anthony Santos
Stanford University
Palo Alto, California

Karsten Schwan
Georgia Tech
Atlanta, Georgia

Shekhar Srikantaiah
Penn State University
College Station, Pennsylvania

Guanying Wang
Virginia Tech
Blacksburg, Virginia

Zhikui Wang
HP Labs
Palo Alto, California

CHAPTER 1

Low-Power, Massively Parallel, Energy-Efficient Supercomputers

IBM Blue Gene Team*

CONTENTS

*IBM: P.V. Allen, C.J. Archer, R.G. Archambault, S. Asaad, J.E. Attinella, J. Balster, J.R. Behun, R.E. Bellofatto, J.R. Bentlage, H.R. Bickford, S.K. Birkholz, M. Blocksome, M.A. Blumrich, A. Boulter, T.C. Brennan, J.J. Brewer, B. Brezzo, A.A. Bright, J.R. Brunheroto, T.A. Budnik, L. Chang, J.D. Chauvin, D. Chen, C.-Y. Cher, G. L.-T. Chiu, T.M. Cipolla, P.W. Coteus, A. Curioni, P. Curtis, K. Davis, M. Deindl, R.H. Dennard, B. Deskin, W. Donegan, J. Doi, M.B. Dombrowa, S.M. Douskey, G. Dozsa, A.E. Eichenberger, D. Eisenmenger, N.A. Eisley, M.R. Ellavsky, S.D. Ellis, K.C. Evans, S.T. Evans, G.A. Fax, A. Ferencz, S. Fetterolf, J.T. Ficke, G. Fiorenza, B.G. Fitch, R.A. Fitch, B.M. Fleischer, W.T. Flynn, T.W. Fox, D.J. Frank, R.L. Franke, S. Frei, M. Fritsch, D.S. Gallo, A Gara, R. Germain, P.R. Germann, M.E. Giampapa, F.P. Giordano, M.P. Good, T.M. Gooding, M.K. Gschwind, J.A. Gunnels, W.H. Haensch, S.A. Hall, M.J. Hamilton, R.A. Haring, J.S. Harveland, P. Heidelberger, T.D. Helvey, D. Hoenicke, R. Hoover, B.J. Hruby, T.A. Inglett, D.J. Iverson, H. Jacobson, G. Janssen, M.J. Jeanson, M.C. Johnson, S.P. Jones, J.N. Judd, K.T. Kaliszewski, R. Kammerer, M. Kapur, F. Kasemkhani, M. Kaufmann, K.H. Kim, B.L. Knudson, S. Koch, M. Kochte, B. Koehler, G.V. Kopcsay, J. Kriegel, E. Kronstadt, S. Kumar, D.E. Lackey, A.P. Lanzetta, C. Lappi, J.A. Lawrence, D.A. Lawson, G.S. Leckband, S. Lee, R.F. Lembach, T.A. Liebsch, D. Littrell, K.C. Lyndgaard, R.W. Lytle, S.H. Mack, C.D. Malone, A. Mamidala, I. Mani, J.A. Marcella, C.M. Marroquin, C.H. Mathiowetz, M.D. Maurice, M.K. McManus, M.G. Megerian, M.P. Mendell, V. Metayer, S.J. Miller, T. Moe, R.K. Montoye, J.H. Moreno, M.B. Mundy, R.G. Musselman, T.E. Musta, I.I. Nair, B.J. Nathanson, Y. Negishi, E. Nelson, M.T. Nelson, C. Nilsen, C.F. Obert, K. O'Brien, A.S. Ohmacht, M. Ohmacht, D. Olson, J.L. Van Oosten, J.P. Orbeck, M.R. Ouellette, M.J. Palmer, J.J. Parker, D.P. Paulsen, K.P. Pfarr, R.A. Rand, M. Rangarajan, J. Ratterman, D.D. Reed, M.T. Repede, D.M. Rickert, T. Roewer, B.S. Rosenburg, M.G. Rosenfield, J.J. Ruedinger, K.D. Ryu, Y. Sabharwal, V. Salapura, D.L. Satterfield, J. Sawada, M. Schaal, P.E. Schardt, M.J. Scheckel, B. Schenck, H.J. Schick, D. Schmunkamp, R.L. Schoen, A.A. Schram, B.A. Schuelke, S. Schwartz, F.W. Sell, G.W. Sellers, R.M. Senger, J.C. Sexton, V.V. Shah, R.H. Sharrar, R. Shearer, J.E. Sheets, E. Shmueli, B. Smith, K.M. Solie, S.A. Strissel, B.D. Steinmacher-Burow, W.M. Stockdell, C. Stunkel, K. Sugavanam, Y. Sugawara, N. Suginaka, T. Takken, A.T. Tauferner, J.L. Thomas, S. Tian, J.A. Tierno, M.R. Tubbs, I. Vo, S. Wahl, C.D. Wait, R.E. Walkup, A.T. Watson, B.B. Winter, B. Wirtz, R.W. Wisniewski, G. Zhang, P.P. Zhao, M.M. Ziegler, C.G. Zoellin, L. Zumbrunnen. Columbia University: N.H. Christ. Columbia University and Riken BNL: C. Kim. Independent: B. Ji. University of Edinburgh: P.A. Boyle. University of Minnesota: S.J. Koester.

1.1 INTRODUCTION

Historically, power considerations have forced the electronics industry to evolve from vacuum tubes, to bipolar device technology, then to NMOS (n-type metal-oxide semiconductor), and finally to CMOS (complementary metal-oxide semiconductor) technology. For the past four decades, steady lithographic advancements have enabled higher integration, leading to exponentially decreasing cost per function, a trend commonly referred to as Moore's law [1]. For the larger part of this period, the semiconductor industry could follow the scaling guidelines developed by Dennard and coworkers [2] to design ever-smaller devices that could operate at ever-higher speeds while keeping power density constant. In recent years, however, fundamental physical limitations have caused CMOS technology to deviate from this ideal device scaling. While trying to maintain speed and density improvements, the industry had to give up on keeping power density constant. Consequently, power dissipation has become a growing concern. Whereas in the past from one device generation to the next we could shrink lithography by a factor $\sqrt{2}$ and expect two times more transistors in the same area while maintaining the same power density, we now find that power nearly doubles for those two times more transistors. Since power is already a constraint across all platforms, from handheld consumer devices to workstations, mainframes, and high-performance computing (HPC) systems [3–7], this is unwelcome news.

CMOS technologies in the 45-nm generation and beyond will require novel solutions to meet the challenge of power efficiency. However, the widespread adoption of parallelism, such as multicore and many-core architectures, in today's computing systems [4–7] creates new opportunities for power/performance optimization. To capitalize on these opportunities, the trade-offs in technology, circuits, and systems design will have to be evaluated from a systems perspective.

Because computing applications span a wide range of power and performance targets, as well as activity factors, the term *low power* can be interpreted in many different ways. This chapter does not focus on low-activity-factor applications, ranging from sensor networks to portable applications, which only require intermittent compute capacity. Power dissipation for such low-activity-factor applications is dominated by standby power. Known techniques that can effectively mitigate standby power in such applications include clock gating, power gating [8], and the adjustment of transistor threshold voltages and gate dielectric thicknesses. Instead, this chapter addresses the more fundamental issue of reducing dissipation in the active mode, which is particularly relevant to applications with high-performance requirements and a high activity factor. We discuss low-power technologies as well as architectural innovations that reduce power consumption in such a context.

In the following analysis, performance and power are considered to be system-level metrics to be optimized. The system is assumed to consist of core logic and its associated cache memory, off-chip main memory, and a power delivery system. While such elements conceptualize the primary components of a parallel supercomputer, many other computer applications are similarly organized. It is important to note that the total power in such systems includes significant contributions from many sources—not simply the processor itself [9].

Section 1.2 analyzes voltage scaling, which enhances power efficiency in the overall system. Section 1.3 presents a practical case study of the IBM Blue Gene® systems. The low-power techniques described here demand an increase in parallelism to compensate for a reduction in operating frequency. Section 1.4 presents the Blue Gene system software and Section 1.5 the Blue Gene applications. In Section 1.5, we demonstrate that the majority of applications for which power efficiency is critical can be effectively parallelized to increase system-level performance in the range of interest. Thus, we can trade off power efficiency, frequency, and parallelism in a systems design.

1.2 VOLTAGE SCALING IN HARDWARE TECHNOLOGIES

1.2.1 Low-Voltage Scaling of Active Devices

For many years, CMOS voltage reduction occurred in conjunction with the reduction of technology dimensions. During this time, the scaling of MOS field-effect transistors (MOSFETs) largely followed the theory outlined in Table 1.1, which was originally proposed in Reference 2. By applying a suitable scale factor to each technology parameter, constant electric fields can be maintained throughout the device as it shrinks in size. Such a strategy protects against short-channel effects and maintains device reliability; more important, it results in reduced circuit delay without increasing power density. While many important advances in transistor technology

TABLE 1.1 Scaling Theory to Maintain Constant Electric Fields in a MOSFET Device

Device or Circuit Parameter	Scaling Factor
Device dimension t_{ox}, L, W	$1/\kappa$
Doping concentration N_a	κ
Voltage V	$1/\kappa$
Current I	$1/\kappa$
Capacitance $\varepsilon A/t$	$1/\kappa$
Delay time/circuit VC/I	$1/\kappa$
Power dissipation/circuit VI	$1/\kappa^2$
Power density VI/A	1

Source: Copyright from L. Chang, D. J. Frank, R. K. Montoye, S. J. Koester, B. L. Ji, P. W. Coteus, R. H. Dennard, and W. Haensch, *IEEE Proc.*, 98(2), 215–236, February 2010. κ is a dimensionless scale factor.

have been made through the years, the basic structure has not changed significantly, and these scaling guidelines, first proposed over 35 years ago, are still relevant.

As CMOS technologies entered the submicron regime, several fundamental issues led to the modification of these scaling rules [10]. In particular, due to nonscalability of the threshold voltage and underlying limits on the subthreshold slope, supply voltage scaling slowed and in recent years has essentially come to a halt. Difficulties in scaling the gate dielectric thickness have also contributed to this trend, as a minimum gate dielectric thickness will have to be maintained to limit leakage power. Consequently, a minimum gate voltage will have to be applied to maintain the electric field and device performance. In addition, as manufacturing variability has a mounting influence on device characteristics, it has been prudent in some cases to raise voltages as a precaution to preserve operating margins. As a consequence, as shown in Figure 1.1, the supply voltage in modern technologies is significantly higher than originally suggested by a scaling theory [11].

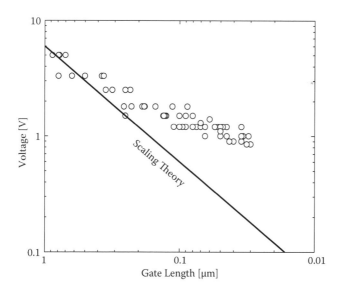

FIGURE 1.1 Scaling trend for power supply voltages in modern CMOS technologies. Due to leakage and variability constraints, voltage levels have deviated significantly from a constant field scaling theory [3]. (Copyright from L. Chang, D. J. Frank, R. K. Montoye, S. J. Koester, B. L. Ji, P. W. Coteus, R. H. Dennard, and W. Haensch, *IEEE Proc.*, 98(2), 215–236, February 2010.)

To first order, power dissipation in the active mode can be expressed as

$$P_{active} = C_{eff} V^2 f + I_{leak} V \tag{1.1}$$

where C_{eff} is the total effective load capacitance of a chip (a more precise definition is given in Section 1.3), V is the operating voltage, f is the operating frequency, and I_{leak} is the total aggregate leakage current of active devices when not being switched. The first term is the dynamic power dissipation due to switching, while the second term is the power consumed by leakage. Since C_{eff} is weakly dependent on voltage, the combined effective voltage dependence of $C_{eff} V^2$ has an exponent closer to 2.5 [12]. Empirically, it has been observed that the maximum operating frequency for a wide variety of circuits is a linear function of voltage in the regime of interest. An expression for frequency can thus be written as

$$f = \alpha(V - V_0) \tag{1.2}$$

where V_0 is the voltage at which the frequency approaches zero ($\sim$0.25 V for modern technologies), and α is a constant that depends on the circuit. This same relation also applies to circuits that are optimized at each voltage, but with somewhat higher V_0 ($\sim$0.3–0.4 V) since low-voltage technologies generally optimize to higher threshold voltages. Putting these equations together yields

$$P_{active} = \alpha C_{eff} V^2 (V - V_0) + I_{leak} V \tag{1.3}$$

While I_{leak} has a strong dependence on voltage, design optimization tends to maintain a consistent ratio between switching and leakage dissipation such that the overall voltage dependence of P_{active} is roughly cubic. Operating voltage is thus clearly the most effective parameter through which power dissipation can be improved. A reduction in voltage, however, limits operating frequencies and inevitably degrades the performance of a given circuit. In accordance with current trends [13], system-level performance can be regained with additional parallel circuit blocks, which linearly adds to power dissipation. Since the superlinear improvements in power due to voltage scaling outweigh the linear increase in power due to parallelism, the end system can see substantial gains in power efficiency.

1.2.2 On-Chip Digital Noise

While active transistors discussed in the previous sections can help to maintain logic performance and memory circuit functionality at low supply voltages, it must also be ensured that signals propagated on properly scaled passive interconnects [13, 14] at these low voltages are resilient to those noise sources relevant to digital circuits. A common initial reaction to voltage scaling is that susceptibility to noise is increased—that at a constant noise level, a reduction in operating voltage could compromise margins. It is important to remember, however, that the noise sources relevant to digital circuit operation will scale with voltage, and that the end impact on circuit functionality may, in fact, improve. It will thus be argued in this section that such on-chip digital noise should not be of concern if voltages continue to be scaled.

Generally, digital CMOS circuits are quite tolerant of noise, but operating margins for some circuits can be small. In particular, dynamic logic can suffer from charge leakage problems, while latches may see degraded setup and hold times. Such voltage noise can be caused by resistive drops, capacitive charge coupling, and inductive transients.

With appropriate consideration of scaling factors, each of these noise sources decreases in importance as voltages are lowered. It should be noted that this discussion neglects mechanisms such as thermal, shot, and $1/f$ noise. While important for analog circuits, such noise is not generally a concern for digital circuits.

Resistive voltage drops as a fraction of the power supply voltage are related to the current drive I of a given device and the characteristic resistance R of the wiring path in question:

$$\frac{\Delta V_R}{V_{DD}} = \frac{IR}{V_{DD}} \propto \frac{(V_{DD} - V_T)^{1.5}}{V_{DD}} \tag{1.4}$$

Since R is not a function of voltage, this expression depends only on I, which, for the purposes of this discussion, can be expressed as a power law function of the gate overdrive voltage [15], where the exponent is assumed here to be about 1.5 for modern CMOS technologies. Since I is a superlinear function of V_{DD}, it scales faster than V_{DD}, and the expression can be seen to decrease with voltage scaling for relevant values of V_{DD} and V_T. Thus, resistive voltage drops become less of a concern with voltage scaling.

Voltage noise due to capacitive coupling occurs when an aggressor of parasitic capacitance C_{agg} switches by a potential difference of V_{DD} and acts on a victim load capacitance C_{vic}:

$$\frac{\Delta V_C}{V_{DD}} = \frac{\frac{C_{agg} V_{DD}}{C_{vic} + C_{agg}}}{V_{DD}} = \frac{C_{agg}}{C_{vic} + C_{agg}} \tag{1.5}$$

The magnitude of this coupling noise as a fraction of V_{DD} is thus determined by the capacitive divider between the victim and aggressor and is not a function of voltage. Thus, voltage noise due to capacitive coupling scales with V_{DD} and does not worsen.

Inductive voltage noise arising from current transients can be calculated as

$$\frac{\Delta V_L}{V_{DD}} = \frac{L\frac{\partial I}{\partial t}}{V_{DD}} \propto \frac{LI}{V_{DD}\tau} \propto \frac{(V_{DD} - V_T)^{1.5}(V_{DD} - V_0)}{V_{DD}} \qquad (1.6)$$

where τ is the characteristic time of such current spikes, which is related to the operating frequency of the circuit in question, which, from Equation 1.2, is linearly dependent on V_{DD}. This overall expression is a superlinear function of V_{DD}, which means that inductive noise scales faster than V_{DD} and thus only improves with voltage scaling.

1.2.3 Power Delivery

Assuming that low-voltage logic and memory solutions are available, and digital noise is contained, the next requirement is to ensure that this low-voltage supply is efficiently and accurately delivered to the chip. Without appropriate consideration of the power delivery system, excessive voltage margins may be needed, which can counteract the gains achieved by successful on-chip voltage scaling. Already today, at about 1-V supplies, the power loss and noise in the path from the external power source to the circuits on a chip can be significant. When voltage is reduced to improve power efficiency at constant performance, total power is lowered, which improves power supply efficiency and stability. However, in scenarios that increase parallelism beyond this point to improve system-level performance, the power supply efficiency and stability may become severely degraded, and advancements in chip packaging or point-of-load power conversion will be required.

In a traditional power delivery system, an off-chip DC-DC (direct-current-to-direct-current) converter normally regulates the supply voltage. The power is then delivered to, and distributed throughout, the chip via a power grid. Nonnegligible power loss occurs in the power delivery network due to Joule heating, which degrades power efficiency. For a system to deliver a power P at a voltage V and total current I through a power delivery line of effective resistance R, the power loss is given by

$$\frac{P_{loss}}{P} = \frac{I^2R}{P} = \frac{(P/V)^2R}{P} = \frac{PR}{V^2} \qquad (1.7)$$

While a reduction in voltage could increase power loss, the corresponding drop in power dissipation levels more than compensates since the dependence of power on voltage in Equation 1.3 is more than cubic. Assuming that parallelism is achieved at the system level and not by growing chip size, it can be assumed that the resistance remains constant. Thus, power delivery efficiency for a fixed design may not degrade but might in fact be improved with voltage scaling.

Supply variation due to sudden load changes can result in a voltage drop, which can be calculated as

$$\frac{\Delta V_L}{V} = \frac{L\frac{dI}{dt}}{V} \propto \frac{\Omega L P}{V^2} \tag{1.8}$$

where L is the inductance of the power distribution network, and Ω is the characteristic frequency over which current loads change (which might be related to the distribution network rather than the voltage-dependent chip operating frequency). As with power loss [15], supply variation scales well with voltage, primarily due to reduced power dissipation levels. The additional dependence on frequency may further suppress supply noise at low voltages—rendering instability a less-critical issue than power delivery efficiency. However, it should be remembered that circuits operating at low voltage may be more sensitive to supply variations. It should also be noted that this issue can also be improved by the addition of more decoupling capacitance.

In the optimizations of Section 1.2.1, voltage scaling will improve power efficiency at constant system performance (i.e., constant $C_{eff} \times f$), which, as stated previously, results in power delivery efficiency and manageable supply variation issues. However, the savings in power due to voltage scaling could instead be used to maximize the number of parallel units for a given power budget, which will improve system performance at constant total power. The number of parallel units could be increased dramatically—constrained only by cost and physical chip size limits. In this constant power scenario, Equations 1.7 and 1.8 indicate that power delivery efficiency and supply variations could worsen significantly with voltage scaling. Thus, new methods of power delivery may be needed.

Vijay Janapa Reddi and colleagues have studied the voltage variation of the Intel Duo processor in detail [16]. In Figure 1.2 from their paper [16], the voltage of an Intel Core™ 2 Duo processor is measured for a variety of workloads. The voltage droops by as much as 9.6% from the nominal, or set point, voltage, which was set 14% above the voltage required for

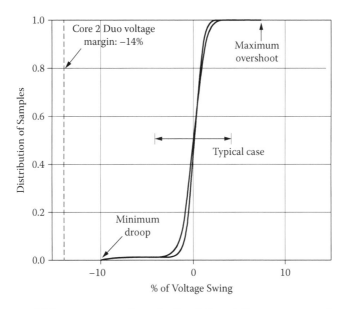

FIGURE 1.2 Voltage measured on the Intel Core 2 Duo processor for a series of workloads. In the worst case, voltage sag was 9.6%, for which the nominal processor voltage was set 14% high. (From V. J. Reddi, S. Kanev, W. Kim, S. Campanoni, M. D. Smith, G.-Y. Wei, and D. Brooks, *IEEE Micro*, pp. 20–26, January/Febuary 2011.)

error-free operation. The authors found a very high correlation between voltage variation and processor stalls, as seen in Figure 1.3. This is easily understood as the voltage undershoot L dI/dt that occurs as the processor moves from the low-power to the high-power state following a stall. Although the authors found some opportunity for software-based predictive algorithms, which may allow a suitable engineered system to operate with voltage closer to the edge of failure, this dramatic voltage variation is predicted by the authors only to worsen in future multicore processor designs.

We expect continued improvements in the efficiency of both alternating current (AC)-to-DC converters and external DC-DC converters. However, unless chip-packaging techniques can be dramatically changed to reduce both resistance and inductance, a new strategy is required to efficiently deliver stable power at low voltage. In modern computer systems, an explosion of different voltages further aggravates the problem. As core voltage falls, other voltages in the form of references, SRAM (static random-access memory) wordline boost voltages, input/output (I/O) driver voltages, receiver

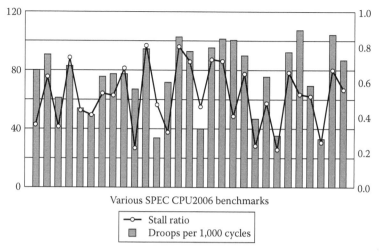

FIGURE 1.3 Voltage droop versus a stall ratio, explained by the authors as a combination of various counters in the processor indication program stall. The correlation between the change from low to high power as indicated by the extent of the stall, and the resultant L di/dt voltage droop, is apparent. (From V. J. Reddi, S. Kanev, W. Kim, S. Campanoni, M. D. Smith, G.-Y. Wei, and D. Brooks, *IEEE Micro*, pp. 20–26, January/Febuary 2011.)

threshold voltages, and the like are proliferating. As a result, a modern computer system can require many different voltages, most of which have modest current demands. This points to an opportunity for on-chip DC-DC voltage conversion. The mix of on-chip and off-chip DC-DC conversion will depend on the area and efficiency of the two solutions. Resistive series regulators are fundamentally limited to low ($\sim$50%) conversion efficiencies due to the inherent resistive divider network and thus are unsuitable for on-chip voltage conversion. Buck converter techniques, utilizing on-chip inductors, are more efficient, but practical implementations have been limited to about 75% due to difficulties in achieving on-chip inductors with high-quality factors. Instead, switched-capacitor circuits may be an effective solution for on-chip voltage conversion. Such circuits traditionally suffer from limitations in efficiency [17], but recent advancements in process technology can potentially enable on-chip conversion efficiencies of more than 90%. The improvement is primarily due to the availability of trench capacitor structures in high-performance CMOS processes. Trenches used for embedded DRAM (dynamic RAM) [18] yield capacitors of very high density with minimal stray parasitics. In addition, with technology scaling, the MOSFETs used as switching devices become quite efficient at the 45-nm node and beyond.

1.2.4 Off-Chip Connections

Combined, solutions to the aforementioned issues can enable low-voltage operation of a chip to improve power efficiency. However, any chip must always communicate with the rest of the system, and this can dissipate significant power. In particular, in many applications, much of the power associated with the memory subsystem can be attributed to such interconnections. Especially for future exascale applications, which will demand extreme memory bandwidth, it is imperative to find solutions that reduce power in off-chip connections.

Off-chip connections that are relatively short, or otherwise operating in a high-quality channel, can be thought of as lossless. Depending on available packaging strategies, such connections can comprise a significant portion of overall I/O power—especially with rising needs in cache bandwidth close to the processor. Without attenuation concerns, the driver and receiver circuits are relatively simple, and the power needed to drive the connection itself can dominate. For these short connections, the active power can be expressed as

$$P_{I/O} = C_{I/O} V^2 f_{\textit{eff}} \qquad (1.9)$$

where $C_{I/O}$ is the interconnect capacitance, V is the operating voltage, and $f_{\textit{eff}}$ is the effective frequency—considering application-dependent activity factors—at which the connection is operated. Clearly, scaling of the output voltage range in these interconnect driver circuits is an effective method by which power could be reduced. The introduction of a locally generated and regulated low-voltage supply can enable a power-efficient, low-voltage driver. On the receiving end, a single-ended sense amplifier, such as enabled by a gated diode device [19], can provide efficient, low-voltage signal recovery. Together, these components can minimize power in low-loss connections. As long as the voltages of all connections are scaled together, signal cross talk can be minimized. For short interconnections that follow Equation 1.9, it may also be possible to reduce interconnect capacitance and the effective frequency of operation. In particular, advanced packaging techniques such as three-dimensional integration via wafer bonding [20] or silicon carriers [21] bring chips closer together, which can eliminate transmission line effects, reduce capacitance, and decrease power as compared with traditional I/O pins and board-level wiring. Ultimately, continued density scaling and single-chip integration will shorten many off-chip connections.

For longer-reach interconnections, which suffer from losses due to high-frequency attenuation, it may be possible to utilize low signal swing to reduce power [22]. But, ultimately, channel quality limits the practicality of such techniques, as the transceiver circuits will tend to dominate total power. Recent work on low-power serial links has focused on equalization techniques [23–25], which may benefit somewhat from the general CMOS voltage-scaling strategies described in this chapter. However, voltage scaling in analog circuits may be limited, and parallelism is likely not a viable solution in this case. Thus, ultimately, optical interconnects [26] will likely be required to achieve significant power reduction in long-range links.

It should be noted that the power associated with off-chip connections can also be dramatically affected by the design and organization of the overall system. For example, since significant energy is consumed in moving data between main memory and the computational engine, the most power-efficient solutions directly attach DRAM chips to the processor chip without intervening address/control or data redrive circuits, hub chips, or other JEDEC (Joint Electron Devices Engineering Council) [27] standardized devices. In addition, the availability of sufficient I/O pins enables operation of off-chip connections at a modest data rate, which allows for source-terminated interconnects and removes the need for far-end bus termination, thus further reducing power.

1.2.5   Cooling

For several reasons, it is desirable to operate computers such that the junction temperature T_j of the CMOS transistors is kept in the range of 20°C to 85°C. The reason for the lower bound is to avoid operating electronic components below the dew point (i.e., the temperature at which water would start to condense out of the atmosphere). The reason for the upper bound is twofold: Repeated excursion between low and high temperature (on and off cycles) can cause thermal cycle fatigue of mechanical connections in the chip package. Also, CMOS devices slow and age faster as temperatures increase. Next-generation HPC systems will contain many optical transceivers, and these have an even more limited thermal range. Light output and aging of optical drivers is strongly temperature dependent. For optical transceivers, it is most desirable to operate at a relatively constant temperature, as close to the dew point as possible.

Within these thermal constraints, we should contemplate how to minimize cost. As large HPC machines are substantial, multimegawatt devices, it is useful to review cooling at the facility level. Unless a large body of water is available nearby, data centers use the atmosphere as a final heat sink. They do so through use of a cooling tower, where return water from the data center is cascaded in a fashion to equilibrate its temperature with the outside air. It is possible in fact for the water temperature leaving a cooling tower and returning to the data center to be several degrees centigrade cooler than the ambient external air. What happens next depends on a choice between air cooling and water cooling. Traditionally, data centers are air cooled, so the water from the cooling tower cycles through a water-to-water heat exchanger, used to cool a closed loop of water, called the chilled water loop. Within the chilled water loop is a water chiller, which reduces the water temperature further, typically to 6°C. At this point, the water pipes need to be insulated to reduce ambient heating of the water and to avoid condensate from the air from forming on the pipes. The chilled water then passes through radiators within computer room air conditioners (CRACs), which use large fans to blow the cooled air beneath a raised floor at a temperature of about 15°C. Alternatively, the air conditioning occurs in larger units within the data center facility, and the cooled air is distributed throughout the data center using air movers and large plenums. Ultimately, the cool air is directed to electronics via local cooling fans within the computer racks. For each watt of power cooled, typically another 0.35 W is used to provide the cooling. In addition, we should consider the cost of purchasing, installing, and servicing the cooling equipment.

A variant of air cooling is hydro-air cooling. Here, the hot air leaving an electronics rack is passed through a radiator, where it is cooled, so that it can be reused for cooling the next rack. Depending on airflow, the water temperature in the radiator, and the heat load, the air temperature leaving the rack could be higher, lower, or the same as the air temperature entering the rack. These systems are advantageous when either the cost of adding additional CRACs is high or if a ready source of water, higher than the 6°C required for CRACs but lower than the exit air temperature, is available for flowing through radiators. Note that the electricity cost of moving air through the CRACs is saved as the fan that directs air through the rack can also drive air through the radiator. A variety of manufacturers makes these radiators, some for individual racks, and some for entire rows of racks.

Their presence and success in the market are appropriate as they can save substantial cost.

An alternative to air cooling is indirect water cooling. It starts in the same manner as air cooling, with a cooling tower and heat exchanger to a closed loop of water. However, this time the closed loop is at 18°C or higher, thus above the dew point of the air in data centers as specified by ASHRAE (American Society of Heating, Refrigerating, and Air-Conditioning Engineers) [28]. This room temperature water enters the computer rack and regulates the temperatures within the rack through contact cooling. Thermal interface materials (TIMs) are used to provide intimate thermal contact between, for example, a copper water pipe and a processor chip or optical transceivers. As it is difficult to cool every component in a rack with indirect water cooling, hybrid systems are used in which most of the heat is exchanged to water and the remaining part to air.

1.3   BLUE GENE HARDWARE

IBM's Blue Gene series of supercomputers—Blue Gene/L [4], Blue Gene/P [5], and Blue Gene/Q [6, 7]—combine many tens of thousands of low-power computing nodes of modest performance to yield massive super-computers that are not only the most power efficient [29], but also the fastest in the world at the time of their introductions in 2004–2011 [30]. Fundamentally based on massive parallelism, Blue Gene systems provide a practical framework within which to discuss the strategies outlined in this chapter. While the Blue Gene systems take some important initial strides toward power efficiency, future systems will likely make more widespread use of the concepts discussed in Section 1.2.

Figure 1.4 shows the Blue Gene/P packaging hierarchy. The packaging hierarchies for Blue Gene/L and /Q are similar. Thirty-two compute cards are packaged onto the next-level board, called the node card. Thirty-two node cards are plugged from both sides into two vertical midplane boards. Thus, 1,024 sockets can be placed and interconnected in a single cabinet. This dense packaging is enabled by the relatively low power of each individual socket. In comparison, a typical rack from other vendors contains about 100 sockets due to power constraints.

The 1,024 compute cards (processor chip with associated DRAM chips) per rack are used as "compute nodes." They are interconnected in a torus topology and run a lightweight operating system (Compute Node Kernel, CNK, described in Section 1.4.1). In addition, a Blue Gene system uses a

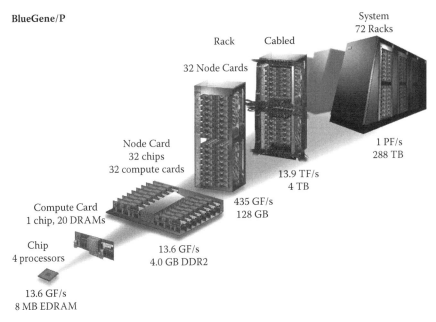

FIGURE 1.4   The packaging hierarchy of the Blue Gene/P system. Each chip has four cores. With 32 chips on a node card and 32 node cards per racks, it has 1,024 chips or 4,096 cores in a rack. GF, gigaflops; TF, teraflops; PF, petaflops, EDRAM, embedded DRAM.

smaller set of compute cards as "I/O nodes" to connect to the file system. I/O nodes are placed either in additional slots on the node cards (Blue Gene/L or/P) or in a separate enclosure (Blue Gene/Q). Logically, these I/O nodes are placed "outside" the torus and are connected to the compute nodes by separate links. The I/O nodes run Linux.

1.3.1   Voltage Scaling

The total compute performance in a rack can be simply described as

$$\text{Performance/Rack} = \text{Performance/Watt} \times \text{Watt/Rack} \qquad (1.10)$$

The watt/rack factor is determined by thermal cooling capabilities. We target about 30 kW (kilowatts) for air-cooled racks and about 100 kW for water-cooled racks. Thus, to maximize the performance of a rack, we will need to optimize the performance/watt factor, that is, the power efficiency. We chose the low-frequency, low-power embedded IBM PowerPC 440 core for Blue Gene/L and the IBM PowerPC 450 core for Blue Gene/P because they had better power efficiency than high-frequency, high-power microprocessors of the time by a factor of about 10, regardless of the manufacturers of the systems. For Blue Gene/Q, we similarly chose the A2 processor, a 64-bit, four-way, multithreaded processor optimized for throughput at relatively low frequency.

The theoretical basis for the superior collective performance of low-power systems was explained in Section 1.2. Any performance metric, such as FLOPS (floating-point operations per second), MIPS (million instructions per second), or SPEC (Standard Performance Evaluation Corporation) benchmarks is linearly proportional to the chip clock frequency. On the other hand, from Equation 1.1, the power consumption of the ith gate on a chip is

$$
\begin{aligned}
P_i &= \text{Switching Power of Gate } i + \text{Leakage Power of Gate } i \\
&= \alpha_i C_i V^2 f + I_i V
\end{aligned}
\tag{1.11}
$$

In the leakage power term, I_i is the leakage current multiplied by supply voltage $V = V_{DD}$. The leakage power becomes increasingly important as technology dimensions shrink. For Blue Gene/L chips, built in 130-nm technology, leakage contributes less than 2% of the system power. For Blue Gene/P, built in 90-nm technology, leakage power is around 8–9% of total power. For Blue Gene/Q, built in 45-nm technology, leakage power is around 14% of total power (in a LINPACK [31] application) but can range up to 20% for chips on the fast end of the manufacturing spread. Operation at the lowest possible voltage for a given technology will minimize the leakage current and thereby leakage power.

In the switching power term, C_i is the load capacitance of the ith gate, and f is the frequency of the fastest on-chip clock. The "switching factor" α_i expresses that not every gate output will switch on every clock cycle at frequency f. The switching factor $\alpha_i = 1$ only for the fastest on-chip clock circuits that both charge and discharge a load capacitance every clock cycle at frequency f, transferring a charge $Q_i = C_i V$ from power supply voltage V to ground. For data circuits, which either charge or discharge a load, at most every clock cycle, $\alpha_i = 0.5$ or less.

The switching power consumed in a chip then is the sum of the power of all switching gates, expressed as

$$
P_{chip} = \Sigma \, (\text{Switching power of gate } i) = C_{eff} V^2 f
\tag{1.12}
$$

where the average switching chip capacitance is given by

$$C_{eff} = \Sigma \alpha_i C_i \qquad (1.13)$$

It is difficult to predict C_{eff} accurately because we seldom know the switching factor α_i of every gate in every cycle. Furthermore, α_i has a dependence on the instantaneous workload of the circuit. To simplify the discussion, we use an averaged value of C_{eff} obtained either from direct measurement or from power modeling tools.

As explained in Section 1.2.1, for a given circuit design, the clock frequency f is roughly proportional to the supply voltage V; thus, the power consumed per chip P_{chip} is proportional to $V^2 f$ or f^3. Thus, it is advantageous in terms of power efficiency to run individual processors (of a given design) at the lowest possible voltage for a given technology, and at the corresponding low frequency, and compensate for the lost performance (proportional to f) by increased parallelization—assuming the intended workloads permit such parallelization. This then is a major part of the Blue Gene design philosophy. Of course, the optimal design point also has to consider the complexities of mechanical component counts and sizes, the power required to communicate between the increased number of processors, the failure rate of those processors, the cost of packaging those processors, and so on. The Blue Gene systems are a complex balance of these factors and many more. Table 1.2 shows an overview of the hardware parameters for the three Blue Gene system generations as related to power efficiency.

Blue Gene systems utilize efficient, voltage-scaled processors combined with system-on-a-chip designs that integrate memory controllers, a network router, and an I/O adapter alongside the processor and local cache. The Blue Gene/L, /P, and /Q processor chips, fabricated in 130-nm, 90-nm, and 45-nm processes, respectively, operate at 700, 850, and 1,600 MHz—frequencies well below that of other processors in similar technologies. Table 1.2 also shows that, as Blue Gene designs progressed, the operating voltage progressively decreased below the nominal voltage for the technology, saving on both active and leakage power. As stated previously, the design philosophy is that system-level parallelism will compensate for individual chip performance.

While current Blue Gene systems employ low voltages that are within technology specifications, future designs may leverage subnominal supply voltages to attain greater power efficiency—eventually driving the need for low-voltage device and memory techniques.

TABLE 1.2 Overview of Blue Gene Hardware Parameters Relating to Power Efficiency

	Blue Gene/L	Blue Gene/P	Blue Gene/Q
Processor type	PPC440	PPC450	PPC A2
SIMD FPU per processor	2 wide	2 wide	4 wide
Threads per processor	1	1	4
Processors per chip	2	4	16 + 1 + 1
Maximum concurrent threads per chip	2	4	64 + 4
Chip technology	Cu-11 (130 nm)	Cu-08 (90 nm)	Cu-45 (45 nm)
Nominal voltage for technology	1.5 V	1.2 V	1.0 V
Blue Gene chip operating voltage under load (for medium-speed chip)	1.5 V	1.1 V	0.85 V
Processor frequency	700 MHz	850 MHz	1600 MHz
Chip peak performance	5.6 GFLOPS	13.6 GFLOPS	204.8 GFLOPS
Main memory	0.5- to 1-GB DDR	2- to 4-GB DDR2	16 GB DDR3
Memory interface bandwidth	5.6 GB/s (16 B wide)	13.6 GB/s (32B wide)	42.6 GB/s (32B wide)
Node power (Chip + DRAMs; Linpack)	~ 16W	~ 23W	~ 68W
Node performance	~ 0.28GFLOPS/W	~ 0.44GFLOPS/W	~ 2.4GFLOPS/W
Nodes per rack	1,024	1,024	1,024
Rack peak performance	5.7 TFLOPS	13.9 TFLOPS	209.7 TFLOPS
System performance (LINPACK)	0.20 GFLOPS/W	0.38 GFLOPS/W	2.1 GFLOPS/W

The bottom row numbers were current entries in the Green500 list June 2012 [29].

1.3.2 Low-Power Floating-Point Units

A common method to increase the FLOPS/watt ratio is to integrate multiple floating-point engines on a chip. This concept can be traced back to the early 1970s when vector processors, such as CDC STAR-100 [32] and Cray-1 [33], were introduced. Although not integrated on the same chip at the time, it was recognized that SIMD (single instruction, multiple data) vector processors can increase the performance without a similar increase in power. SIMD floating-point units (FPUs) are also being used extensively in recent HPC systems, such as IBM Cell Engines [34], Intel's Polaris chip [35], and GPUs (graphics processor units) [36].

A modern FPU consumes typically from 0.1 W to a few watts [37, 38], depending on many factors, such as frequency, width of the data path (single precision versus double precision), and so on. However, increasing FPUs from 1 to 8 units incurs a relatively minor increase in the overall chip power. Hence, Blue Gene/L and Blue/P used a two-way SIMD double-precision FPU for each processor core, doubling peak performance and increasing power efficiency at the same time. Blue Gene/Q went further in the same direction and incorporates a four-way SIMD double-precision FPU for each processor core.

Projecting into the future, both voltage-scaling and low-power SIMD FPUs will continue to be the foundation of power-efficient supercomputer designs.

1.3.3 Power Delivery

In a massive parallel system, generating, monitoring, and preserving the requisite supply voltages can be challenging—due to the sheer number of voltages required and the stringent requirements on supply robustness and redundancy. With increased parallelism and voltage scaling, these issues will become more severe in future-generation systems.

To facilitate efficiency in chip-to-chip communications and to enable cycle-reproducible operation, a single system clock is distributed to all compute nodes in a Blue Gene system. Since each processor chip must run at the same frequency, tailored supply voltages are used to minimize power in the presence of process variations. In manufacturing, chips are speed sorted into groups. The 32-way node cards are populated homogeneously with chips of a given speed sort bin, and the power supplies on the node card are adjusted—with the fastest chips (which also have the most leakage current) running at a lower voltage and the slowest chips running at a higher voltage. This approach minimizes both active and leakage power.

With a multitude of supplies operating at high current levels and energy densities, redundancy, supply stability, and infrared (IR) with resistive losses must be carefully addressed. To ensure high system reliability, supply voltages are provided by distributed point-of-load converters with $N + 1$ or $N + 2$ redundancy. While a supply failure may create a large transient response, the frequent near-instantaneous changes of up to 40% in processor power due to synchronous clocks and power-gating techniques also create large voltage transients. The power converter loop response and recovery from a supply failure must be fast enough so that the voltage droop does not fall below the minimum voltage required to run the processor. To ensure this, the nominal voltage is raised high enough to cover the worst-case voltage droop and thus adds to the nominal power dissipation. Power planes on the circuit cards are designed to minimize these droops and match the voltage delivered to each processor chip. Processor nodes far from the power supply have additional conduction paths in other circuit card layers to reduce resistance while parallel connections may be removed from other nodes. As a result, supply equipotentials are created at each processor chip location. The number of power planes on a circuit card is fixed, however, such that as extra power supply voltages are added (as needed for DRAM and I/O), distribution losses inevitably worsen.

On Blue Gene/Q, the final stage of DC-DC voltage conversion is performed directly on the system planar. In future systems, as discussed in Section 1.2.3, the final stage of DC-DC voltage conversion will ideally be performed on the processor chip itself. By delivering higher voltages closer to the chip, supply stability and resistive losses can be improved, which will be especially important as operating voltages are further scaled down. In addition, local voltage generation and regulation could significantly simplify the power delivery system by reducing the number of supplies needed.

1.3.4 Off-Chip Connections

As discussed in Section 1.2.4, the power associated with moving data between processor and memory is minimized if the memory controller is located directly on the processor chip. In Blue Gene systems, each processor chip contains a wide interface (16B wide for Blue Gene/L; 32B wide for Blue Gene/P and /Q) to directly attached memory (see Table 1.2).

Source-terminated I/O cells that are impedance matched to the transmission lines between the processor and DRAM eliminate the need for other data line termination. The power dissipated in these lines follows Equation 1.9, where the capacitance is minimized by placing memory immediately adjacent to the processor chip. Using variable-voltage I/O cells and corresponding power supplies, low-voltage memory can be introduced as it becomes available.

Going forward, processor-to-memory bus frequencies will increase. Thus, the power dissipated in the connections between the processor and external memory will increase unless the interconnect capacitance can be reduced to compensate for increasing bus frequencies. By utilizing high I/O count DRAM stacks, based on through-silicon vias or other high-wiring-density interconnect media, large amounts of external DRAM can be placed adjacent to the processor with greatly reduced wiring path lengths.

The longest off-chip connections in a Blue Gene system are used for communication between racks. In Blue Gene/L and /P these links, which can be up to 8 m in length, are managed by electrical cables using differential signaling at up to 3.4 Gb/s (gigabits per second). In Blue Gene/Q, the rack-to-rack communication links are optical fiber connections. In optical connections, attenuation is substantially less than in electrical connections, which means that links of up to 100 m will operate with little or no more power than links of a few meters—a characteristic that may lead to new system design paradigms.

1.3.5 Cooling

Blue Gene/L is entirely air cooled. At 25 kW/rack, it is a relatively high-power machine but still coolable in most data centers without difficulty. In traditional data centers, the airflow enters the machine room proper through perforations in the raised floor, is typically pulled through electronics racks from a "cold aisle" in front of a row of racks to a "hot aisle" in back of a row of racks, and then somehow makes its way back to the CRAC units on the perimeter of the raised floor or, via exhaust vents, to a facility-level air conditioning plant. Indeed, many data centers must leave vast expanses of valuable machine room floor unoccupied to allow enough perforations for the volume of air required for high-power HPC machines.

Blue Gene/L improves on the efficiency of the airflow. As shown in Figure 1.5, cold air is drawn directly from the raised floor into a tapered side

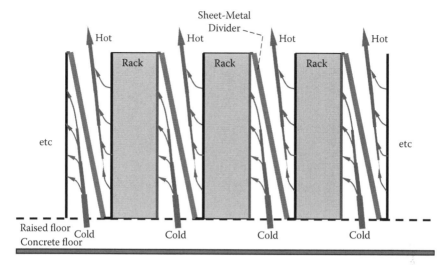

FIGURE 1.5 Airflow through a row of Blue Gene/L or /P racks, showing complementary tapered plenums.

plenum. A wall of fans pulls the air across the rack and exhausts into a second plenum, which directs the warm air to the ceiling. A ceiling plenum returns the air to the air conditioners. Only the bulk power supplies on top of the rack are cooled with air in the machine room proper, and even this air is drawn from the aisle and directed upward, permitting all aisles to be "cold aisles." Thus, the Blue Gene/L solution allows a compact design, with aisle width limited only by the needs of people access, further reducing cost. Further, the unique side-to-side airflow permits a midplane design without air perforations, so that a maximum number of circuit cards can be connected together without electrical cables, thus reducing power and cost.

Blue Gene/P increased the rack power to 32 kW, which is more challenging for air cooling. However, the side-to-side airflow of Blue Gene allows the straightforward introduction of hydro-air cooling, in the form of radiators between racks. These radiators have sufficient cooling capability to cool the air leaving a rack back to its entrance air temperature. A cost analysis comparing air cooling with CRACs to hydro-air cooling for Blue Gene/P, including estimates for equipment purchase as well as electricity cost of $0.10/kWh, resulted in about $7,500 savings per rack over a 5-year lifetime. This analysis includes the purchase and power of a commercial water conditioner to raise the chilled water temperature to above the dew point before entering the rack radiator. If a ready supply of 18°C closed-loop water is available, the cost savings more than doubles. The largest installed

Blue Gene/P system, a 72-rack, 1-PetaFLOPS machine in Jülich, Germany, is cooled via the hydro-air technique and a facility-level water conditioning unit.

Per Table 1.2, Blue Gene/Q has a power efficiency of about 2.1 GigaFLOPS/W. With an aggregate peak performance of 209.7 TeraFLOPS/rack, a highly efficient application could consume nearly 100 kW/rack. This exceeds the capabilities of air cooling. On each 32-way node card assembly within a Blue Gene/Q compute rack, all major components (compute cards, power supplies, link chips, and optics) are clamped to cold rails that are cooled with water that flows through a single, continuous serpentine copper tube. As discussed, the inlet water temperature is about 18°C (above dew point), keeping the optical transceivers well below 50°C and the compute chips below 85°C. Other system components (bulk power supplies, I/O drawers) are still air cooled.

1.4 SYSTEM SOFTWARE

1.4.1 Overview

Blue Gene installations can scale to large node counts. For example, the Jülich Blue Gene/P installation mentioned counts 72-Ki nodes (288-Ki processors), and the Sequoia Blue Gene/Q installation counts 96-Ki nodes (over 1.5-Mi processors). Such large systems pose many challenges for the design of the system software:

1. **Scalability:** System software that needs to scale to hundreds of thousands of cores requires solid software design and strict adherence to the principles embodied by the design.

2. **Performance:** System software should not be the performance-limiting factor. We maintained the Blue Gene design philosophy that the machine speed should be limited by hardware constraints.

3. **Functionality:** Although Blue Gene systems are primarily aimed at HPC applications, we continue to look for opportunities for non-HPC applications to utilize the computing power of Blue Gene.

4. **Power Efficiency:** Machines of this size consume a tremendous amount of power. We designed system software that helped understand and reduce the amount of power used by applications.

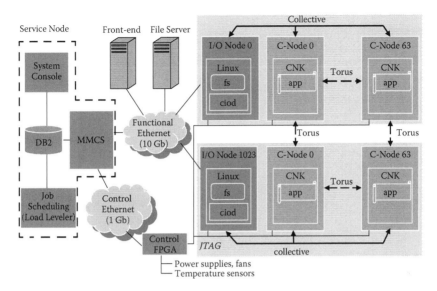

FIGURE 1.6 Blue Gene system software organization. Shown are the software components for compute nodes, I/O nodes, control system, front-end and file servers, and the interconnect networks.

As shown schematically in Figure 1.6, a Blue Gene system consists of several hardware components and interconnection networks, with corresponding software [39].

- Compute nodes form the bulk of a Blue Gene system. Compute nodes are interconnected with high-bandwidth networks: a 3D torus network and a separate collective network for both Blue Gene/L and/P; a unified 5D torus network for Blue Gene/Q. Compute nodes run the actual computational applications, with operating system services supplied by a lightweight Compute Node Kernel (CNK). CNK provides a familiar POSIX (portable operating system interface) interface. Notable is that file system operations are function shipped; they are off-loaded to the I/O nodes that interface with the file system. Another notable feature is that CNK only supports a single-user application, and once the user application is running, it normally only provides the services requested by the application. This makes application run characteristics predictable, with little operating system "noise."

- The I/O nodes are Blue Gene compute chips that are located "on the face" of a Blue Gene machine. They do not participate in the torus network that defines the "compute volume" of the machine but are connected to the compute nodes via separate links. I/O nodes run Linux, mount the file systems, and interface to external computers called "front-end nodes." It is notable that I/O nodes run CIOD (control and input/output daemon), which provides control (job launching, signaling, termination) and file system services for the compute nodes.

- The control system software, MMCS (Midplane Management and Control System), runs on a separate "host" computer (service node) external to the Blue Gene racks. MMCS is responsible for various tasks, such as allocating and booting partitions of Blue Gene nodes, scheduling jobs to run on those partitions, and other resource management activities. In addition, MMCS threads monitor the hardware on defined intervals, generally around 5 min. For boot, monitor, and debug functions, MMCS communicates with a Blue Gene system via a private control Ethernet network, which has a tree topology. The leaves of this control network are control field-programmable gate arrays (FPGAs) located on each node board, which interface with power supplies, temperature sensors, and via a JTAG (Joint Test Action Group) (IEEE 1149.1) interface, with each Blue Gene compute, I/O or link chip.

In the next sections, we describe the system monitoring capability of the MMCS and its ability to record job history. We show how they can be combined into a powerful tool for analyzing and understanding the power consumption on a Blue Gene machine. We then present four different operational phases provided by the system software and describe their power optimizations.

1.4.2 System Monitoring

The Blue Gene control system, MMCS, continuously monitors the hardware status and thereby has several important roles in helping to monitor and understand power consumption on Blue Gene machines.

1. MMCS reads the status of all of the hardware on the machine. This includes node cards, link cards, service cards, fan modules, power modules on cards, individual nodes, and bulk power modules. Combined, there are over 200 unique pieces of "environmental" data

(voltages, currents, temperatures, status flags, fan speeds) gathered from the hardware and stored in a database.

2. When anomalies occur in the hardware readings, MMCS sends warnings and error messages into the Blue Gene database that can be monitored by an administrator. For example, if a piece of hardware is over the warning threshold for temperature, not only will the temperature be recorded along with the location and timestamp, but also a warning event will be issued. This allows an administrator to just look at warning and error events and not have to look at all of the recorded environmental data.

3. MMCS can take hardware off-line should the problem be severe enough. For example, if there are repeated failures of a bulk power module, the hardware monitor component can take a rack off-line and prevent jobs from running on that rack until after it is serviced.

Since voltage and current are being recorded for the bulk power modules as well as for all the node cards, link cards, and service cards on the machine, a large amount of data is available to understand the power behavior of the machine. The control system provides a simple and intuitive way to query this database and analyze and display the data. Thus, power usage can be analyzed for a particular partition or for the entire machine, either at a desired point in time or for the duration of a job or collection of jobs.

1.4.3   Job History

The Blue Gene control system software also stores the details of the jobs that have run on the machine. For each job, the control system stores the executable, the arguments, the start time, the end time, the exit status, the user name, and many other attributes of the job into a database, including the number of Blue Gene nodes that the job occupied and the specific location of those nodes. Part of this extensive job history is useful in determining power usage. By combining the specific location of the hardware used, the starting and ending times of the job, and the detailed power usage measurements for the hardware, the power utilization can be determined on a per job basis.

This means that users can answer questions like, "Which jobs use the most power?" or "In a particular month, how much power did each job use?" An example query showing the power and flexibility of the database approach follows:

```
SELECT JobId, NodesUsed, timestampdiff(4, char(entrydate-starttime))
as RuntimeInMinutes, decimal(timestampdiff(4,char(entrydate-
starttime))/60.0 * avg(output0voltage * output0current),8,2) *
(nodesused/1024) * 9/1000 as KWBulk
     FROM tbgpjob_history a, (select blockid,substr(bpid,1,3) as
rack from tbgpbpblockmap group by blockid, substr(bpid,1,3) having
count(*) <> 1) as x, bgpbulkpowerenvironment c
     WHERE x.blockid = a.blockid and c.time>starttime and
c.time<entrydate and substr(c.location,1,3) = x.rack and date
(entrydate) > = '10/01/2009' and date(entrydate) < = '10/31/2009'
     GROUP BY jobid,nodesused, starttime,entrydate order by 4 desc
```

This query takes the voltage and current output as measured from the bulk power modules and correlates them to jobs that have run on those racks of hardware. It then computes the power usage by capturing the data from the nine bulk power modules for the duration of a job and, in this example, confines the results to jobs that ran in October 2009. Therefore, the result of the query is a listing of the jobs that ran during the month ordered by the kilowatts of power that they used.

These data are gathered and recorded continuously without intervention or setup needed on the part of the user or administrator. Thus, information on historical power usage, the temperature of the racks, or fan speeds (or for Blue Gene/Q, water temperatures) for any job that was run is readily available.

1.4.4 Operational Phases

The Blue Gene system software provides four different operational phases. Power is optimized in each of these phases. In this section, we present a description of each phase and then provide details on that phase and on the power optimizations that occur during that phase. The four phases are as follows:

Deallocated: the period when the node software is not functional;

Booting: the period in which node software is loaded and configured on the hardware;

Allocated: the period when the node software is booted but is not running a job;

Running: the period in which an active job is running on the node.

Before describing the power optimizations for each phase, we present a general workflow for job execution on a Blue Gene system. The first step is to identify the specific nodes that will be used for the job. On Blue Gene, a collection of nodes on which a job can be executed is called a block. A user or system administrator can use Blue Gene commands to define a block or may use scheduling software that interacts with the Blue Gene software to define a block. A block is defined by a number of nodes and, optionally, the shape of the torus. Thus, a collection of compute nodes and I/O nodes is identified and given a unique name on the system. Once identified, the block must be booted before it can run any jobs. The default state for a block is deallocated, meaning that the block has been defined, but it has not been booted, so the nodes are in reset and not running any software. When a block is to be booted, the control system must first validate that all hardware identified by the block is available, and that no overlapping blocks are already booted on the hardware. Once this check has been performed, the control system starts the nodes and begins loading the kernels onto the nodes. This process is the booting phase. The booting phase is completed when the CNK is loaded onto all compute nodes, the Linux kernel is loaded onto all I/O nodes, every I/O node has reported that Linux has successfully initialized, and all file system mounts have completed. When all nodes have reported success, the block goes to an allocated status. This means that it is ready to run jobs, but no jobs are running yet. The CIOD daemon running on each I/O node is waiting to be contacted. The final step is when a job is submitted, either directly by a user or via a scheduler. This occurs via a command called mpirun, which contacts the I/O node daemons and tells them which binary executable to load from the mounted file system and start it, along with details such as arguments to the job and environment variables. When all I/O nodes have successfully launched the job on their connected compute nodes, the block is in running status.

Next, we describe the power optimizations that occur during each of the four operational phases.

1.4.4.1   Deallocated

Traditionally, an HPC system is booted once, and compute jobs are launched on behalf of specific users. On a Blue Gene system, the control system can deallocate unused sections of the compute nodes, called a block. When a block is not allocated, the control system holds the compute nodes in reset, which is an extremely low-power state. All logic inside the chip is held at a steady state, with no latch switching to consume power. When the compute nodes are needed again, the control system must release the chips from reset, which requires a boot of the CNK on the compute nodes. This is not too onerous due to the lightweight nature of CNK.

1.4.4.2 Booting

The control system makes several power optimizations during a boot. One of the first steps in booting a block (i.e., a set of compute and I/O nodes) is enabling the clocks to each processor chip. Typically, the chip logic is divided into multiple local clock domains. Some of the logic is activated optionally based on the location and function of the node. For example, each BLC (Blue Gene/L Compute) and BPC (Blue Gene/P Compute) chip contains an Ethernet MAC (Media Access Control) function. However, only chips deployed as I/O nodes actually use the Ethernet functionality, to connect to front-end nodes and the file system. The control system will therefore only activate the Ethernet clock domain on I/O nodes. The BQC (BlueGene/Q Compute) chip similarly has a PCIe (peripheral component interconnect express) function that is only activated on I/O nodes.

Because the Ethernet or PCIe I/O functions are not being clocked on compute nodes, which form the bulk of the system, this selective clocking policy saves power.

During the booting of a block, the control system will broadcast boot-up and configuration commands to all the chips in the block, including clock domain startup commands. Accordingly, the compute chips on a node card, typically in the same block, will activate their clock domains nearly simultaneously. Such a sudden step up in current demand may cause a pronounced voltage droop across the chip. To mitigate this effect, the control system will stagger the activation commands for the various on-chip clock domains. This "soft start" approach smoothes the step rises in current and thereby reduces the demands on the power distribution system.

1.4.4.3 Allocated

When a block is booted but has no job running on the compute nodes, the operating system enters an idle state, waiting for the next application to run. With a general-purpose operating system, even in the idle state there will be some activity on the node due to periodic interrupts and background dae-mons, making it hard to minimize power consumption. With a lightweight operating system such as CNK, the system knows it is waiting for just one or two things, the next command from the control system or from the I/O node, and it can wait for those in as efficient a manner as possible. On Blue Gene/L and /P, the waiting takes the form of a polling loop in which both code and data reside in the processor's L1 cache, effectively quiescing the rest of the memory system. As the memory system is one of the large power draws, this technique reduces overall power consumption. On Blue Gene/Q, CNK goes further and parks all but one of the 68 hardware threads in *wait* instructions, where they consume no execution unit or memory system resources at all. (The one thread that remains active, polling the L1 cache during this phase, is a CNK node controller thread.) When a command arrives from the I/O node, the Blue Gene/Q *wakeup unit* will break one thread out of its wait instruction. This thread will then control the download and start of the new application.

1.4.4.4 Running

Once a job is running on the Blue Gene nodes, the temperature and power consumption of the node can vary significantly within phases of the application and across different applications. There are two main voltage domains on the node, one voltage domain for the compute chip and another voltage domain for the off-chip memory. CNK has the ability to throttle power consumption on a specific voltage domain using either reactive or proactive power management. There are several techniques for slowing the processor domain; CNK relies on a periodic interrupt and spin wait to accomplish this task. To throttle the off-chip memory voltage domain, the compute chip's memory controller can be placed into a special mode that adds some stall cycles to certain memory accesses, thereby giving the node board's power distribution system time to recover from the previous burst of memory activity. This technique allows the average power demand to be smoothed.

Reactive power management is activated when the node card tells CNK that either the temperature or power consumption is exceeding a preset threshold. CNK will throttle both voltage domains once the power supply warning is detected. Since each node acts independently, CNK communicates this reactive mode to all of the other nodes executing the job. The purpose of this communication is so that the nodes can coordinate their power-throttling activities. This scheduling is important to avoid performance noise due to each node randomly reducing performance. When a threshold-exceeded event occurs, the nodes will only have a few milliseconds to respond before the power supplies can no longer deliver the current power levels. This tight timing constraint means that CNK (rather than the control system) must react to the power management request. Once the reactive mode has been activated, the control system is informed via a RAS (reliability, availability, serviceability) message.

Proactive power management is similar to reactive power management except that the user conveys that the job should start with proactive power throttling. The user can also specify which voltage domain to target along with various settings on intensity. Proactive power management is intended for applications that are known to exceed a certain power profile. This allows the application to maintain control and specify when or where power management occurs without invoking reactive power management. Although the application may specify proactive mode, the power supplies could still hit the reactive power management thresholds. If this occurs, then the reactive power management settings will override the proactive settings. When proactive mode is activated, the control system is informed via a RAS message.

1.4.5   Low-Power Implications on Software

As mentioned, a focus on power efficiency resulted in the hardware design using relatively simple, low-frequency processor cores, as the system is optimized for collective performance, as opposed to single-thread performance. For example, each core has a simpler design (e.g., less branch prediction capability) than other modern high-performance processors. Thus, wherever system software is adversely affected by single-thread performance, optimization work needs to be done. For example, it required some work to achieve the targeted I/O bandwidth, to activate OpenMP® (Open Multi-Processing) threads and to enable "hybrid" programming models, such as MPI (message-passing interface) with OpenMP. Most affected, however, is the messaging software stack.

We provide a few examples of the optimizations in the messaging software stack. One way to maximize performance within the design resource constraints is to take advantage of the DMA (direct memory access) facility in the torus network logic. The DMA is especially helpful when moving large amounts of data. Instead of tying up a processor core for many cycles while moving data, the messaging software only spends a few processor cycles setting up the DMA. Then, the DMA engine takes over and moves the data efficiently between the torus network logic and memory without further processor involvement. This minimizes the messaging overhead, and therefore the power, in the Blue Gene/P and BlueGene/Q torus networks.

Additional optimizations are required in the messaging software stack to meet the performance provided by the hardware. The messaging code is written to help the compiler better utilize the instruction cache and minimize branches by using "likely_if()" and "unlikely_if()" pragmas. A likely_if() branch tells the compiler the branch is usually going to be taken; therefore, it would be good to preload instructions along that path and to set up the branch prediction such that the branch is assumed. The messaging code also eliminates as many branches as possible and optimizes the expected-case paths for latency. In addition, certain operations, such as accessing the DMA hardware or addressing virtual-to-physical address mappings, can be expensive due to system calls or long latency access paths to device configuration registers. As this information changes infrequently, a copy is cached in the application's address space for fast access.

Significant effort was spent reducing overhead for collective calls. As the node counts increase, the costs of doing many point-to-point send/receive calls increases. To minimize this overhead, the component collective messaging interface (CCMI) was developed. CCMI is based on the concept of multisends. Each processor is involved only once in a given multisend call. Each collective has a data movement "schedule" that defines data flow and operations on the data. The schedule is then "executed" by the node in the multisend call. This helps absorb overhead that would normally be present in a collective operation in which each node calls send or receive multiple times.

1.5 APPLICATIONS

This chapter described how power and packaging considerations have driven the Blue Gene design point, in particular how power-efficient supercomputers can be built from moderate-performance processor chips, compensating for individual processor performance by increasing system-level parallelism. This section shows that this approach is actually valid, that is, that a significant class of HPC applications scales well to very large processor counts.

The first major Blue Gene/L system installation occurred in 2005 at Lawrence Livermore National Laboratories (LLNL). This system included 64 racks, had 131,072 individual compute cores, and delivered a peak performance of 367 TeraFLOPs. When it was installed, it took the number one spot on the international list of top 500 supercomputers [30]. In that same year, the runner-up systems on the top 500 list had core counts on the order of 10,000 or less. Blue Gene/L's core counts thus represented a very significant increase over what, at the time, was the norm.

Today, managing massive concurrency in parallel applications is well understood, and there are many systems installed with core counts in the 50,000+ range. At the time of Blue Gene's first installation, however, from the application programmer's point of view, the order of magnitude increase in core count was a potential barrier to adoption of Blue Gene systems. In its design, however, Blue Gene had a significant number of innovations to support efficient large-scale parallelism for applications. A number of application teams completed considerable preparation work during the Blue Gene development phase to prepare applications that could take advantage of what was, at the time, a unique architecture.

In the years since Blue Gene was first introduced, a wide variety of applications and application domains has successfully been ported to and are in production use on Blue Gene systems around the world. Application domains span from modeling of materials and proteins at the nanoscale; to computational fluid dynamics and structural mechanics in engineering applications; and to systems biology, whole-organ simulation, and gene sequence analysis in the life sciences domain.

As examples of successful applications that make optimal (and thereby the most power-efficient) use of the machine, we discuss the four applications that won separate Gordon Bell awards in the period 2005 to 2007. The Gordon Bell awards are given each year for the highest-performance applications submitted to the annual Supercomputing Conference—the International Conference for High Performance Computing, Networking, Storage, and Analysis.

- In 2005, the Gordon Bell prizewinner was a classical molecular dynamics simulation of metal solidification, which achieved over 100 TeraFLOPs of sustained performance on the LLNL Blue Gene system [40]. The core algorithm for this simulation requires a very large number of independent force calculations for the individual atoms in the simulation. This calculation in itself is easily parallelized, so that performance scaling becomes dependent on managing the communications requirements, in this case making optimal use of the special integrated torus network built into Blue Gene. Thus, performance scaling was achieved by careful domain decomposition of the

TABLE 1.3   Strong Scaling Performance for the Molecular Dynamics Code on Blue Gene/L as Reported in Reference 40.

Processor Cores	Atoms	Atoms/Core	Elapsed Time (s)	Scaling Fraction
1,024	2,048,000	2,000	461.72	1.00
2,048	2,048,000	1,000	245.68	0.94
4,096	2,048,000	500	139.87	0.83
8,192	2,048,000	250	70.24	0.82
16,384	2,048,000	125	40.61	0.71
32,768	2,048,000	62.5	26.63	0.54
131,072	2,048,000	15.625	18.52	0.19

This table shows how overall wall clock time is reduced for a fixed-size problem as processors are added. For 1,024 processor cores, this simulation assigns 2,000 atoms per core, so that for 131,072 cores, there are 15.6 atoms per core on average. Scaling is not perfect, but out to 32,768 processors, scaling is about 50%. Even at 131,072 processors, meaningful performance gains are achieved.

problem across the high Blue Gene node count. Table 1.3 demonstrates the strong scaling performance for one specific data set described in Reference 40. This data set modeled the solidification of a system of 2 million uranium atoms. The table shows solution time decreases for this fixed problem size as the processor count increases. Ideally, one would like to see solution time decreasing proportional to the increase in processors used. The table shows reasonable scaling from 1,024 to 32,768 processors. At 131,072 processors, scaling has fallen off but still shows decreasing solution time. A variation of this application code also won the 2007 Gordon Bell prize [41].

- In 2006, Blue Gene won two Gordon Bell prizes. The first was for a quantum molecular dynamics simulation of metals, which achieved 207 TeraFLOPs on the LLNL system [42]. This performance equated to 56% of peak, enabled by two key Blue Gene capabilities. The first of these was the bandwidth match between memory system and processor cores (see Table 1.2), enabling high performance on dense matrix operations. The second of these was again the integrated torus network, which, with careful decomposition of the problem, delivers scalable communications.

- The second 2006 Gordon Bell winner was for a subnuclear chemistry application [43], which is a weak scaling problem (i.e., the problem size grows with the size of the machine). Table 1.4 shows perfect weak scaling from 1,024 to 131,072 cores for two different solution methods. Again, the integrated torus network was key to delivering this result.

TABLE 1.4   Application Scaling Performance for QCD (Quantum Chromo Dynamics) on Blue Gene/L as measured in the context of Reference 43.

Racks	Processor Cores	Dirac Operator Sustained Performance per Core (%)	Conjugate Gradient Inverter Sustained Performance per Core (%)
0.5	1,024	19.2	18.7
1	2,048	19.2	18.7
2	4,096	19.3	18.6
4	8,192	19.2	18.6
8	16,384	19.3	18.6
16	32,768	19.3	18.5
20	40,960	19.3	18.5
64	131,072	19.2	18.5

This table shows almost perfect performance scaling from 1,024 to 131,072 processor cores. Data for two different kernels are presented. These kernels include the Dirac operator kernel, which is a sparse matrix vector operation, and a full conjugate gradient inverter, which inverts the Dirac operator. The table shows sustained performance (as a percentage of peak performance) of both kernels in weak scaling mode, where the size of the vectors involved is scaled with the number of cores, thus keeping the work per core constant.

- Recently, Blue Gene also demonstrated the most energy efficient solution for very-large-scale data uncertainty quantification: More than 9 TB of data could be validated by estimating the inverse of the data covariance matrix in less than 20 min. This application ran on the full 72 Racks Blue Gene/P at the Jülich Research Center performing at 73% of theoretical peak [44]. The simultaneous presence of the torus and integrated collective network was instrumental to the efficient implementation of this algorithm.

As demonstrated by these examples, it is possible to build scalable, robust HPC systems that can provide significant application performance at concurrency levels one to two orders of magnitude greater than previously proven. The very tight integration of networking, memory, and processors has been proven to enable application scalability to very high processor core counts. With such architectural optimizations in place, and software that takes advantage of them, these examples demonstrate the feasibility of driving power efficiency into HPC with massively parallel systems based on tight integration of low-power, moderate-performance processor cores.

1.6 CONCLUSIONS

In conclusion, power and power efficiency will be playing an increasingly important role in HPC. The key to power efficiency is a combination of architecture and technology choices. On the technology side, low voltage is critical to power efficiency. On the architecture side, however, simply taking an existing high-performance optimized design and scaling it to lower voltage will not achieve optimal power efficiency. Instead, to achieve an optimal system design, resources such as memory bandwidth, and interconnect bandwidth should also use appropriate low-power technology choices. This integrated approach has been extremely successful in the Blue Gene program over multiple generations of machines.

As we look forward, exponentially higher levels of concurrency will be necessary to achieve the desired (and forecasted) performance levels. As the performance grows, better power efficiency will be essential. The lessons learned in the Blue Gene program will be leveraged, but additional techniques will be necessary. Architectural innovation will be a critical enabler of usable future systems as voltage scaling will be increasingly challenging, and as foreseeable technology advances may provide only minor relief in terms of power efficiency.

Low-level power management of systems will become commonplace, but in the end, systems must deliver performance and be highly usable. As users continue to migrate to programming models that are consistent with high levels of single-node parallelism, new architecture options will open up that can more easily achieve both exceptional sustained price performance and power efficiency.

Supercomputers will continue to be the venue that first experiences difficult computing issues and will be the place where solutions are first implemented and widely utilized. Recently, the focus on power has also taken on an environmental component, transcending simple cost considerations. The concept of responsible computing is emerging in the supercomputing community, and this nearly equates to power-efficient computing.

ACKNOWLEDGMENTS

This work has benefitted from cooperation of many individuals at IBM Research (Yorktown Heights, NY) and at the IBM Systems and Technology Division (Rochester, MN; East Fishkill, NY; Burlington, VT; and Raleigh, NC). The Blue Gene/L, Blue Gene/P, and Blue Gene/Q projects have been supported and partially funded by Argonne National Laboratory and the Lawrence Livermore National Laboratory, on behalf of the U.S. Department of Energy, under Lawrence Livermore National Laboratory subcontracts no. B517552 and B554331.

Parts of the first three sections of this chapter originally appeared in *IEEE Proceedings*, 98(2), 215–236, 2010, and we are indebted to those authors [3].

REFERENCES

1. G. E. Moore, Progress in digital integrated electronics, *IEDM Tech. Dig.*, 11–13, 1975.
2. R. H. Dennard, F. H. Gaensslen, H. N. Yu, V. L. Rideout, E. Bassous, and A. R. LeBlanc, Design of ion-implanted MOSFETs with very small physical dimensions, *IEEE J. Solid-State Circuits*, SC-9, 256–268, October 1974.
3. L. Chang, D. J. Frank, R. K. Montoye, S. J. Koester, B. L. Ji, P. W. Coteus, R. H. Dennard, and W. Haensch, Practical strategies for performance-efficient computing, *IEEE Proc.*, 98(2), 215–236, February 2010.
4. A. Gara, M. Blumrich, D. Chen, G. L.-T. Chiu, P. Coteus, P. Coteus, M. E. Giampapa, R. A. Haring, P. Heidelberger, D. Hoenicke, G. V. Kopcsay, T. A. Liebsch, M. Ohmacht, B. D. Steinmacher-Burow, T. Takken, and P. Vranas, Overview of the Blue Gene/L system architecture, *IBM J. Res. Dev.*, 49(2/3), 195–212, 2005.
5. G. Almasi, S. Asaad, R. E. Bellofatto, H. R. Bickford, M. A. Blumrich, B. Brezzo, A. A. Bright, J. R. Brunheroto, J. G. Castaños, D. Chen, G. Chiu, P. Coteus, M. Dombrowa, G. Dozsa, A. E. Eichenberger, A. Gara, M. Giampapa, F. Giordano, J. A. Gunnels, S. A. Hall, R. Haring, P. Heidelberger, D. Hoenicke, M. Kochte, G. Kopcsay, S. Kumar, A. Lanzetta, D. Lieber, B. J. Nathanson, K. O'Brien, A. S. Ohmacht, M. Ohmacht, R. A. Rand, V. Salapura, J. C. Sexton, B. D. Steinmacher-Burow, C. Stunkel, T. Takken, S. Tian, B. M. Trager, R. B. Tremaine, P. Vranas, R. E. Walkup, M. E. Wazlowski, S. Winograd, R. W. Wisniewski, P. Wu, D. R. Busche, S. M. Douskey, M. R. Ellavsky, W. T. Flynn, P. R. Germann, M. J. Hamilton, L. Hehenberger, B. J. Hruby, M. J. Jeanson, F. Kasemkhani, R. F. Lembach, T. A. Liebsch, K. C. Lyndgaard, R. W. Lytle, J. A. Marcella, C. M. Marroquin, C. H. Mathiowetz, M. D. Maurice, E. Nelson, D. M. Rickert, G. W. Sellers, J. E. Sheets, S. A. Strissel, C. D. Wait, B. B. Winter, C. J. Wood, L. M. Zumbrunnen, M. Rangarajan, P. V. Allen, C. J. Archer, M. Blocksome, T. A. Budnik, S. D. Ellis, M. P. Good, T. M. Gooding, T. A. Inglett, K. T. Kaliszewski, B. L. Knudson, C. Lappi, G. S. Leckband, S. Lee, M. G. Megerian, S. J. Miller, M. B. Mundy, R. G. Musselman, T. E. Musta, M. T. Nelson, C. F. Obert, J. L. Van Oosten, J. P. Orbeck, J. J. Parker, R. J. Poole, H. L. Rodakowski, D. D. Reed, J. J. Scheel, F. W. Sell, R. M. Shok, K. M. Solie, G. G. Stewart, W. M. Stockdell, A. T. Tauferner, J. Thomas, R. H. Sharrar, S. Schwartz, D. L. Satterfield, J. D. Chauvin, E. Shmueli, R. G. Archambault, A. R. Martin, M. P. Mendell, G. Zhang, P. P. Zhao, I. Mani, R. Nair, R. D.

Bendale, A. Curioni, Y. Sabharwal, J. Doi, and Y. Negishi, Overview of the IBM Blue Gene/P project, The Blue Gene Team, including G. Chiu, *IBM J. Res. Dev.*, 52(1/2), 199–220, 2008.

6. R. A. Haring, M. Ohmacht, T. W. Fox, M. K. Gschwind, D. L. Satterfield, K. Sugavanam, P. W. Coteus, P. Heidelberger, M. A. Blumrich, R. W. Wisniewski, A. Gara, G. L.-T. Chiu, P. A. Boyle, N. H. Christ, and C. Kim, The IBM Blue Gene/Q compute chip, *IEEE Micro*, 32(2), 48–60, 2012.

7. Blue Gene/Q: Sequoia and Mira, in *Contemporary HPC Architectures*, ed. J. S. Vetter, Taylor and Francis, Boca Raton, FL, April 2013.

8. J. Tschanz, S. Narendra, Y. Ye, B. Bloechel, S. Borkar, and V. De, Dynamic-sleep transistor and body bias for active leakage power control of microprocessors, *IEEE Int. Solid-State Circuits Conf. (ISSCC)*, 102–103, February 2003.

9. K. Rajamani, C. Lefurgy, S. Ghiasi, J. Rubio, H. Hanson, and T. Keller, Power management solutions for computer systems and datacenters, Tutorial, International Symposium on High-Performance Computer Architecture, February 2008.

10. G. Baccarani, M. R. Wordeman, and R. H. Dennard, Generalized scaling theory and its application to a 1/4 micrometer MOSFET design, *IEEE Trans. Electron Devices*, ED-31, 452–462, April 1984.

11. E. J. Nowak, Maintaining the benefits of CMOS scaling when scaling bogs down, *IBM J. Res. Dev.*, 46, 169–180, March/May 2002.

12. R. H. Dennard, J. Cai, and A. Kumar, A perspective on today's scaling challenges and possible future directions, *Solid-State Electron.*, 51, 518–525, April 2007.

13. M. Horowitz and W. Dally, How scaling will change processor architecture, *IEEE Int. Solid-State Circuits Conf. (ISSCC)*, 132–133, February 2004.

14. J. Meindl, Low power microelectronics: Retrospect and prospect, *Proc. IEEE*, 83, 619–635, April 1995.

15. T. Sakurai and A. R. Newton, Alpha-power law MOSFET model and its applications to CMOS inverter delay and other formulas, *IEEE J. Solid-State Circuits*, 25, 584–594, April 1990.

16. V. J. Reddi, S. Kanev, W. Kim, S. Campanoni, M. D. Smith, G.-Y. Wei, and D. Brooks, Voltage noise in production processors, *IEEE Micro*, 20–26, January/February 2011.

17. G. Patounakis, Y. W. Li, and K. L. Shepard, A fully integrated on-chip DC-DC conversion and power management system, *IEEE J. Solid-State Circuits*, 39, 443–451, March 2004.

18. G. Wang, K. Cheng, H. Ho, J. Faltermeier, W. Kong, H. Kim, J. Cai, C. Tanner, K. McStay, K. Balasubramanyam, C. Pei, L. Ninomiya, X. Li, K. Winstel, D. Dobuzinsky, M. Naeem, R. Zhang, R. Deschner, M. J. Brodsky, S. Allen, J. Yates, Y. Feng, P. Marchetti, C. Noris, D. Casarotto, J. Benedict, A. Kniffin, D. Parise, B. Khan, J. Barth, P. Parries, T. Kirihata, J. Norum, and S. S. Iyer, A 0.127 μm^2 high performance 65nm SOI-based embedded DRAM for on-processor applications, *IEDM Tech. Dig.*, 1–4, December 2006.

19. W. K. Luk and R. H. Dennard, Gated diode amplifiers, *IEEE Trans. Circuits Syst. II Express Briefs*, 52(5), 266–270, May 2005.

20. S. J. Koester, A. M. Young, R. R. Yu, S. Purushothaman, K.-N. Chen, D. C. La Tulipe, Jr., N. Rana, L. Shi, M. R. Wordeman, and E. J. Sprogis, Wafer-level 3D integration technology, *IBM J. Res. Dev.*, 52, 583–597, 2008.

21. P. S. Andry, C. K. Tsang, B. C. Webb, E. J. Sprogis, S. L. Wright, B. Dang, and D. G. Manzer, Fabrication and characterization of robust through-silicon vias for silicon-carrier applications, *IBM J. Res. Dev.*, 52, 571–581, 2008.

22. R. Palmer, J. Poulton, W. J. Dally, J. Eyles, A. M. Fuller, T. Greer, M. Horowitz, M. Kellam, F. Quan, and F. Zarkeshvari, A 14mW 6.25Gb/s transceiver in 90nm CMOS for serial chip-to-chip communications, *IEEE Int. Solid-State Circuits Conf. (ISSCC)*, 440–441, February 2007.

23. Y. Liu, B. Kim, T. O. Dickson, J. F. Bulzacchelli, and D. J. Friedman, A 10 Gb/s compact low-power serial I/O with DFE-IIR equalization in 65nm CMOS, *IEEE Int. Solid-State Circuits Conf. (ISSCC)*, 182–183, February 2009.

24. Y. Hidaka, W. Gai, T. Horie, J. H. Jiang, Y. Koyanagi, and H. Osone, A 4-channel 10.3Gb/s backplane transceiver macro with 35dB equalizer and sign-based zero-forcing adaptive control, *IEEE Int. Solid-State Circuits Conf. (ISSCC)*, 188–189, February 2009.

25. J. F. Bulzacchelli, T. O. Dickson, Z. T. Deniz, H. A. Ainspan, B. D. Parker, M. P. Beakes, S. V. Rylov, and D. J. Friedman, A 78mW 11.1Gb/s 5-tap DFE receiver with digitally calibrated current-integrating summers in 65nm CMOS, *IEEE Int. Solid-State Circuits Conf. (ISSCC)*, 368–369, February 2009.

26. D. A. B. Miller, Rationale and challenges for optical interconnects to electronic chips, *Proc. IEEE*, 88, 728–749, June 2000.

27. Joint Electron Devices Engineering Council home page. http://www.jedec.org/.

28. American Society of Heating, Refrigerating, and Air-Conditioning Engineers (ASHRAE) home page. http://www.ashrae.org/.

29. The Green500 List, November 2007–2013. http://www.green500.org/lists.php.

30. The TOP500 List, November 2004–November 2013. http://www.top500.org/.

31. J. Dongarra, P. Luszczek, and A. Petitet, The LINPACK benchmark: Past, present and future, *Concurr. Comput.*, 15(9), 803–820, 2003.

32. R. G. Hintz, and P. Tate, Control data STAR-100 processor design, *COMPCON, IEEE*, pp. 1–4, September 1972.

33. R. M. Russell, The Cray-1 computer system, *Commun. ACM*, 21(1), 63–72, January 1978.

34. IBM to build world's first cell broadband engine based supercomputer, IBM Press announcement, September 6, 2006, http://www.ibm.com/press/us/en/pressrelease/20210.wss.

35. S. R. Vangal, J. Howard, G. Ruhl, S. Dighe, H. Wilson, J. Tschanz, D. Finan, A. Singh, T. Jacob, S. Jain, V. Erraguntla, C. Roberts, Y. Hoskote, N. Borkar, S. Borkar, An 80- Tile Sub-100-W TeraFLOPS Processor in 65-nm

CMOS, *IEEE Journal of Solid-State Circuits,* 43(1), 29–41, Jan. 2008, http://dx.doi.org/10.1109/JSSC.2007.910957

36. J. D. Owens, D. Luebke, N. Govindaraju, M. Harris, J. Krüger, A. E. Lefohn, and T. Purcell. A survey of general-purpose computation on graphics hardware, *Comput. Graph. Forum,* 26(1), 80–113, 2007.

37. J. Balfour, W. J. Dally, D. Black-Schaffer, V. Parikh, and J.-S. Park, An energy-efficient processor architecture for embedded systems, *IEEE Comput. Arch. Lett.,* 7(1), 29–32, 2008.

38. W. J. Dally, J. Balfour, D. Black-Schaffer, J. Chen, R. C. Harting, V. Parikh, J.-S. Park, and D. Sheffield, Efficient embedded computing, *IEEE Comput.,* 27–32, July 2008.

39. J. E. Moreira, G. Almasi, C. Archer, R. Bellofatto, P. Bergner, J. R. Brunheroto, M. Brutman, J. G. Castanos, P. G. Crumley, M. Gupta, T. Inglett, D. Lieber, D. Limpert, P. McCarthy, M. Megerian, M. Mendell, M. Mundy, D. Reed, R. K. Sahoo, A. Sanomiya, R. Shok, B. Smith, and G. G. Stewart, Blue Gene/L programming and operating environment, *IBM J. Res. Dev.,* 49(2/3), 367–376, 2005.

40. F. H. Streitz, J. N. Glosli, M. V. Patel, B. Chan, R. K. Yates, B. R. de Supinski, J. Sexton, and J. A. Gunnels. 100+ TFlop solidification simulations on Blue Gene/L, Gordon Bell Prize, Supercomputing 2005. http://sc05.supercomputing.org/schedule/pdf/pap307.pdf.

41. J. Glosli, K. Caspersen, D. Richards, R. Rudd, F. Streitz (Lawrence Livermore National Laboratory), and J. Gunnels (IBM, Inc.), Extending stability beyond CPU-millennium: Micron-scale atomistic simulation of Kelvin-Helmholtz instability, in *Proceedings of the 2007 ACM/IEEE conference on Supercomputing* (SC '07). ACM, New York, NY, USA, Article 58. http://doi.acm.org/10.1145/1362622.1362700

42. F. Gygi, E. W. Draeger, M. Schulz, B. R. De Supinski, J. A. Gunnels, V. Austel, J. C. Sexton, F. Franchetti, S. Kral, C. Ueberhuber, and J. Lorenz, Large-scale electronic structure calculations of high-Z metals on the Blue Gene/L platform, in *Proceedings of the 2006 ACM/IEEE conference on Supercomputing* (SC '06). ACM, New York, NY, USA, Article 45. http://doi.acm.org/10.1145/1188455.1188502

43. P. M. Vranas, G. Bhanot, M. Blumrich, D. Chen, A. Gara, P. Heidelberger, V. Salapura, and J. C. Sexton, The Blue Gene/L supercomputer and quantum chromodynamics, in *Proceedings of the 2006 ACM/IEEE conference on Supercomputing* (SC '06). ACM, New York, NY, USA, Article 50. http://doi.acm.org/10.1145/1188455.1188507.

44. C. Bekas, A. Curioni, and I. Fedulova, Low cost high performance uncertainty quantification, *Proceedings of the 2nd Workshop on High Performance Computational Finance,* ACM, New York, 2009; http://doi.acm.org/10.1145/1645413.1645421.

CHAPTER 2

Compiler-Driven Energy Efficiency

Mahmut Kandemir and Shekhar Srikantaiah

CONTENTS

2.1   INTRODUCTION

Conserving energy at the cost of minimal performance degradation is an important goal in high-performance computing systems. Energy conservation can be achieved in the hardware by various mechanisms, like dynamic voltage and frequency scaling [1], state transitions among various power states, or temporarily turning off some resources when they are not being used. Hardware enhancements for energy optimization have been reasonably well studied in the past. However, energy optimizations cannot leverage these hardware mechanisms without adequate support from the system software. Software like the compiler, operating system, and run-time systems can act as *enablers* for these energy-conserving mechanisms to be used effectively. Without these enablers, hardware optimizations alone cannot accrue the predicted energy benefits. In particular, the compiler is capable of automatically analyzing program behavior, data access patterns, and resource requirements of complex application programs.

An energy-aware compiler can reshape a program either explicitly, by inserting instructions to turn off resources that are not used, or implicitly, by increasing the idle periods between two usages of the resource. A compiler framework is the ideal platform for performing such energy-aware transformations automatically. For example, the efficacy of previously proposed energy optimization techniques like dynamic disk speed modulation (dynamic rotations per minute, DRPM) [2] depend on the availability of idle disk usage periods. If we execute unmodified code or code optimized for high performance, it has been observed [3] that the opportunity for finding such idle periods is diminished. Energy-aware compiler optimization such as described in Reference 4 can be used to increase the idle periods, thereby facilitating energy conservation. In this light, it is important to understand the nuances of compiling for high performance versus compiling for energy conservation.

2.1.1 Compiling for Performance versus Energy

High-performance systems make extensive use of optimizing compilers. Most optimizing compilers are focused on improving the performance of applications. In some ways, optimizing the performance of a program reduces the total amount of "work done" and thereby reduces its energy consumption. But, this is not always true. Energy-aware compiler optimizations are generally harder than and significantly different from performance optimizations because all parts of the program execution contribute to its overall energy consumption. In performance optimizations, only the *critical path* of a program execution needs to be optimized, whereas a part of the program not on the critical path, like speculative execution or I/O (input/output), prefetches may contribute significantly to the energy consumption of the program. Therefore, it is crucial to give a fresh thought to compiling high-performance applications with both performance and energy as first-order metrics for optimization.

2.1.2 Avenues for Energy-Aware Compiling

Looking ahead into future high-performance computing systems, it is evident that they will be characterized not only by their raw performance but also by diverse communication capabilities and power/energy management. In particular, two trends are going to be responsible for the increasing energy demands of future high-performance applications.

The first trend is the enormous increase in data-intensive applications. Peta-scale high-performance computing has enabled scientists to tackle very large and computationally challenging problems, such as those found in the scientific computing domain. This in turn helps advancement of scientific discovery at a faster pace. However, this is also leading to a tremendous increase in the amount of data used by these programs. The amount of data is doubling every 18–24 months as a result. All these data must be stored, organized, and processed in real time to be made useful. Therefore, the I/O subsystem is turning out to be a major energy consumer in these systems. Add to it the mechanical nature of disks, which leads to higher amounts of energy consumption. Solid-state disks (SSDs) are being introduced to ameliorate this problem. However, there are other issues with SSDs, like reliability, that need to be addressed before their widespread use. Therefore, disks are guaranteed to constitute high performance in the future for another decade or so.

The second trend is the simultaneous increase in both on-chip communication (in the case of chip multiprocessors, CMPs) and off-chip communication with massively parallel applications parallelized for modern peta-scale computing platforms. In the future, with many cores on chip, it has been forecast that the on-chip communication will be in the form of a network on a chip (NoC). Power consumed by these NoCs and off-chip communication fabric is raising major concerns that need to be addressed to make them feasible. Fortunately, compilers with the knowledge of synchronization and data sharing among application threads are best suited for performing optimizations to reduce both on-chip and off-chip communication.

A considerable fraction of the energy consumed in modern high-performance systems can be attributed to two major components in addition to others in high-performance systems:

- The *I/O subsystem* performing I/O on terabytes of data is a major consumer of energy.

- On-chip communication (CMPs) and off-chip communication are increasing, with higher levels of parallelism being targeted, leading to increased energy consumption.

Recent advances in energy-efficient compilation for high-performance computing systems have been focused on these two aspects. This chapter elaborates on compiler-directed energy optimizations in both the I/O subsystem and the NoCs.

2.2 ENERGY-AWARE I/O OPTIMIZATIONS

Disk systems of the large, high-performance machines are known to contribute to a significant fraction of the overall power budget. Motivated by this observation, recent studies (e.g., [5], [3]) specifically focused on the disk system and proposed energy-saving strategies. These efforts are either hardware based (e.g., reducing disk speed if the associated latency can be tolerated by the application) or software based (e.g., restructuring code for taking best advantage of available low-power capabilities provided by the disk system). In particular, from the compiler's perspective, there are two major parameters that can be tuned for conserving energy:

- The code structure of the application program, and

- The disk layout of data.

Both approaches have been investigated recently. A detailed description of the techniques are discussed next.

2.2.1 Modifying Application Code

There are a number of compilation techniques used to statically analyze program sections and discover disk reads/writes that may be responsible for causing higher energy consumption or to discover opportunities to reduce energy consumption. Son et al. proposed several compiler-based code transformation techniques to conserve disk energy consumption. First, they studied a compiler technique that inserts explicit disk power management calls in source codes of scientific applications [3]. The idea is that a compiler can extract how disks are traversed during execution time using the application source code along with the file-level striping information. By inserting explicit power management calls (e.g., spin_up and spin_down) in the application code, one can eliminate (to a large extent) the performance penalty that would normally be incurred by reactive disk power management schemes. Second, they revisited conventional loop distribution and iteration space tiling techniques from an energy perspective. To achieve the best energy savings without slowing performance much, they showed that both code and underlying disk layout must be considered at the same time. In another paper [6], the same authors described a compiler approach to reduce disk power consumption in the presence of parallel disk systems. To increase disk idleness, the proposed technique schedules the code fragments assigned to a number of processors according to the disk access patterns extracted by an optimizing compiler, which captures both intraprocessor and interprocessor disk reuses.

Modern high-performance applications are becoming increasingly dynamic in terms of their computation patterns as well as data access patterns. Due to the complexities of systems and applications and their high energy consumptions, it is very important to address research issues and develop dynamic techniques to scale I/O in the right proportions. Compiler optimizations are no exceptions to this. Recent research has also studied incorporating a dynamic compilation framework, including a set of powerful I/O optimizations designed to minimize execution cycles and energy consumption. The designed framework generates results that are competitive with hand-optimized codes in terms of energy consumption. Before delving into the details of each of these frameworks, it is important to emphasize that some of the techniques are dependent on profiling programs to collect program characteristics. Such profile-based techniques are more applicable in high-performance computing systems where application characteristics are dominated by input-independent behavior more than the input-dependent ones. Moreover, most of the techniques described here try to avoid collecting input-dependent behavior from profiles to the extent possible. We now discuss both a static compilation framework and a dynamic compilation framework for modifying application code in an energy-aware manner.

```
for i=0 to N − 1 {
   for j=0 to M − 1 {
      ... V₁[i][j] ...
   }
}
```

FIGURE 2.1   Example of code fragment.

2.2.1.1   Static Compilation Framework

As previously mentioned, a compiler can effectively determine at compile time (statically) many important I/O system parameters, like disk speeds, data layouts, and disk data prefetching distance of applications. One such integrated static compilation framework has been proposed [4].

Consider the example code fragment shown in Figure 2.1. The code fragment given in this figure accesses a two-dimensional disk-resident array, named V_1, using a loop nest constructed from two loops. For illustrative purposes, V_1 is assumed to be striped over four disks with a stripe size of S (see Figure 2.2), and all four disks in question are assumed to be running at 12,000 RPM. As depicted in Figure 2.3a if we do not apply any prefetching (PF_i), every access to the first data element in each block incurs an access R_i to the disk system. In this example, we assume that it takes T_d cycles to complete a disk access when the rotational speed of disks is 12,000 RPM. After T_d cycles elapse, the requested data block D_i is ready; thus, the computation on that block can proceed.

Since most scientific applications are amenable to static analysis of access patterns, they can be extracted and reshaped by an optimizing compiler. We can use the software prefetching algorithm proposed by Brown et al. [7] to hide disk I/O stall time and reduce overall execution latency. The code fragment after applying I/O prefetching is given in Figure 2.4. Software prefetching generates a prolog, a steady state, and an epilog from each original loop nest. The prefetch distance d (i.e., the number of iterations ahead of which the disk I/O needs to be initiated to hide I/O latency), can be calculated as

$$d = \left\lceil \frac{T_d}{s + T_{pf}} \right\rceil, \tag{2.1}$$

where T_d is the estimated I/O latency (in cycles) to prefetch one block, T_{pf} is the overhead (again in terms of cycles) of executing a prefetch

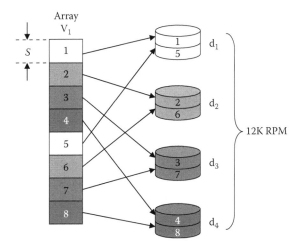

FIGURE 2.2   Disk layout of V_1.

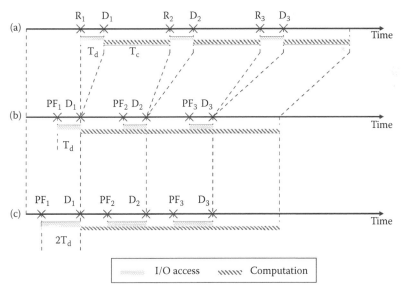

FIGURE 2.3   Comparison of I/O timings. (a) Original code without prefetching. (b) Prefetching to high-speed disks. (c) Prefetching to low-speed disks. T_d is the disk I/O time for a single block data.

```
for i=0 to N − 1 {
  PF (&V₁[i][0]); /* prolog */

  for jj=0 to M − 1 − d, step b₁ { /* steady-state */
    PF (&V₁[i][j + d]);
    for j=jj to jj + b₁ {
      ... V₁[i][j] ...
    }
  }
  for j=M-d to M − 1 { /* epilog */
    ... V₁[i][j] ...
  }
}
```

FIGURE 2.4   Code with prefetching; d is the prefetch distance.

instruction, and s is the number of cycles in the shortest path through the loop body. Once the prefetch distance d is calculated, we can then stripe mine the loop nest to make explicit the point at which the prefetch instruction is to be inserted. The result of this transformation for the example is given in Figure 2.3. In this example, d iterations of j loop are assumed to be required to hide I/O latency, and $b1$ is the strip size used for stripe mining.

Up to this point, we discussed software prefetching as a technique that can be used to hide disk I/O latency, specifically hiding T_d, as proposed in the literature. However, if we examine the components of disk I/O time, we can see that T_d is composed of seek time, rotational latency, transfer time, and controller overhead. Since in modern disk drives the controller overhead is negligible compared to the other three values, we can see that T_d is almost directly proportional to the disk rotation speed. However, it has been shown by prior research that the disk power consumption is quadratically proportional to the disk rotational speed [2]. This suggests that one can take an approach to conserve disk energy by storing array data in low-speed disks, such as a disk running at less than 12,000 RPM (in this example), and by eliminating the increased I/O latency using software prefetching with an increased prefetch distance. That is, one can save disk energy by increasing prefetch distance and reducing disk speed at the same time.

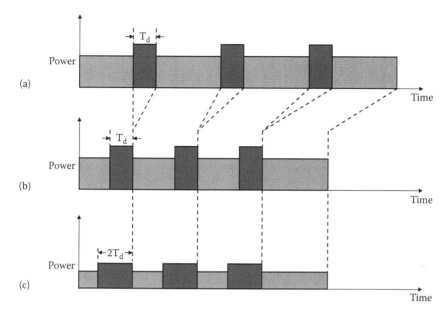

FIGURE 2.5 Comparison of disk power states. (a) Original code without prefetching. (b) Prefetching to high-speed disks. (c) Prefetching to low-speed disks. T_d is the disk I/O time for a single block data.

Figures 2.4 and 2.5 show how prefetching to high-speed disks and low-speed disks affects I/O timing and disk power consumption. In this example, the rotational speed of the low-speed disks is assumed to be 6,000 RPM (i.e., half of the maximum speed possible). Consequently, the time it takes to complete a disk access is doubled, i.e., it is now $2T_d$. One can see from Figures 2.4b and 2.4c that we can hide the latency of low-speed disks by issuing the prefetch early enough. Specifically, since the I/O latency is doubled from T_d to $2T_d$, the prefetch distance d is also doubled based on Equation 2.1. On the other hand, the energy consumption profiles after applying prefetching with different prefetch distances are depicted in Figure 2.5. Figure 2.5a shows the power profile throughout the program execution time when no prefetching is employed. Note that it has been assumed that the disk drive can be placed in either active mode when servicing I/O request or idle mode when the disk is not used. Therefore, the disk is in the active mode during T_d when there is a request being processed. For the remaining time, the disk is placed into the idle mode. Figures 2.5b and 2.5c show how the prefetching affects the power consumption profile of a disk. If we apply prefetching using high-speed disks, we can conserve disk energy consumption by the amount of reduced execution time. In this case, the energy savings come from the reductions in the total disk idle time. In comparison, as shown in Figure 2.5c, if the data are stored in low-speed disks and we apply prefetching, we can reduce disk energy consumption further by cutting the energy consumption in the active periods as well.

It should be noted that we may not be able to take advantage of low-speed disks for all disk-resident arrays because using low-speed disks entails longer prefetch distances, which may not be very appropriate for a loop nest whose iteration count is not sufficient for hiding such a long I/O latency. Therefore, one needs to be careful when selecting the disk speeds to employ. Also, since we focus on large scientific programs that consist of multiple loop nests, it is possible that the determined disk speed for a particular array in one loop nest may not be appropriate for another loop nest that manipulates the same array (by accessing the same set of disks). Consequently, selecting prefetching distance and disk speeds depends on the disk layout of data as well as the data access patterns exhibited by the application code being optimized. Because of this, these parameters should be considered together. We now describe an integrated compiler framework that determines these parameters together.

Before describing the algorithm, let us first define a few important mathematical concepts. An array-based, loop-intensive program $\mathcal{P}$ that consists of s loop nests can be represented as

$$\mathcal{P} = (\mathcal{L}_1, \mathcal{L}_2, \dots, \mathcal{L}_s),$$

where $\mathcal{L}_i$ ($i = 1, 2, \dots, s$) is the ith loop nest in program $\mathcal{P}$. A loop nest $\mathcal{L}_i$ can further be in the following form:[1]

> $\mathcal{L}_i$: for $i_1 = l_1$ to u_1, step b_1
> 　　for $i_2 = l_2$ to u_2, step b_2
> 　　　$\dots$
> 　　　　for $i_k = l_k$ to u_k, step b_k
> 　　　　{loop body}

can be represented as

$$\mathcal{L}_i = \text{for } \vec{I} \in [\vec{L}_i, \vec{U}_i], \text{ step } \vec{b} \ \langle a_1(\vec{I}), a_2(\vec{I}), \dots, a_m(\vec{I}) \rangle,$$

---

[1] If $\mathcal{L}_i$ is not perfectly nested, one can use techniques such as code sinking [8] to make it perfectly nested.

where $\vec{I}$ is the iteration vector; $\vec{L} = (l_1, l_2, \ldots, l_k)^T$ and $\vec{U} = (u_1, u_2, \ldots, u_k)^T$ are the lower and upper bound vectors, respectively; $\vec{b} = (b_1, b_2, \ldots, b_k)^T$ is the loop step vector; and $a_j(\vec{I})$ ($j = 1, 2, \ldots, m$) is the jth array reference in the body of loop nest $\mathcal{L}_i$. While executing, loop nest $\mathcal{L}_i$ is assumed to access n arrays, $V_1, V_2, \ldots, V_n$. Let $\mathcal{V}$ represent a set comprised of these n arrays. The array element accessed by $a_j(\vec{I})$ ($j = 1, 2, \ldots, m$) can be represented as $V_i[\vec{F}(\vec{I})]$ ($i = 1, 2, \ldots, n$, $j = 1, 2, \ldots, m$), where V_i is the name of the array, and function $\vec{F}$ maps iteration vector $\vec{I}$ to a vector of subscripts for array V_i. Specifically, $\vec{F}(\vec{I})$, which maps k loop iterators into d array indices, where k is the depth of the loop nest and d is the dimensionality of the array, can be defined as

$$\vec{F}(\vec{I}) = M\vec{I} + \vec{o},$$

where M is a $d \times k$ matrix (called the access matrix), $\vec{I}$ is k-element iteration vector, and $\vec{o}$ is an offset vector [9].

Let us also assume that the multispeed disks provide l different rotational speeds: $RPM = (1, 2, \ldots, l)$, where 1 represents the lowest disk speed and l corresponds to the highest disk speed available. Last, we define the disk layout for each array V_i using a triplet of the following form:

$$(start\_disk, stripe\_factor, stripe\_size),$$

where start_disk is the first disk where the file striping starts from, stripe_factor is the number of disks being used for striping, and stripe_size is the unit size of each file stripe residing on each disk. The compiler approach determines a prefetch distance for each array access in the application code, a rotational speed for each disk in the storage system, and a data layout for each disk-resident array manipulated by the application.

To exploit low-speed disks using prefetching to save energy, the prefetching algorithm described in Reference 4 needs to analyze the data locality exhibited by each loop nest $\mathcal{L}_i$ in program $\mathcal{P}$. To extract the spatial reuse vector space, we simply replace the last row in M with zeros to create a reduced access matrix M_S and solve for nullspace of M_S, which gives us $span(M_S)$. After determining the temporal/spatial reuse vector spaces, we next choose the set of innermost loop iterators that can exploit reuse. This is called *localized iteration space* [9]. This space captures only those loops for which data reuse can result in data locality. To translate the obtained reuses to locality, we need to take into account the loop iteration count and available memory capacity. Since the loop bounds are assumed to be known at the compile time (if not, we make use of available profile data), one can determine the set of innermost loops whose accessed data fit in the main memory capacity. Data locality is then captured by intersecting the reuse vector space with the localized iteration space, where both are represented by vector space notation. These steps to analyze reuse and data locality exhibited in the given programs are fundamentally unaltered from those developed in the context of conventional I/O prefetching [7]. However, to support prefetching to multispeed disks for reducing disk power consumption, we need to be careful in selecting prefetch distance for every disk-resident array reference, as discussed in detail in the following material.

Using the obtained vector space representation of data locality exhibited by each loop $\mathcal{L}_i$, we next determine prefetch distance (d value in Equation 2.1) for each array reference ($V_i[\vec{F}(\vec{I})]$) made by the loop body of nest $\mathcal{L}_i$. Note that, once the d value is calculated and reference $V_i[\vec{F}(\vec{I})]$ is found to have spatial locality on the ith loop, the ith loop is strip mined, where $1 \leq i \leq k$ and k is the depth of loop nest. Generally, prefetches are software pipelined around this ith loop that changes the value of the array-indexing function $V_i[\vec{F}(\vec{I})]$. This chosen loop is called the *pipeline loop*. As mentioned in the previous section, if we put the data in low-speed disks, the prefetch distance linearly increases with respect to disk I/O time (i.e., the T_d value in Equation 2.1), while power consumption is quadratically reduced by the amount of disk speed scaling [2]. Therefore, we need to tune the prefetch distance based on the disk speed, as explained next.

In the first step of the energy-aware prefetching algorithm, the disk speeds that will provide the maximum energy savings for each array in the application code are determined. In processing an array reference, we consider all possible disk speeds (RPM levels) and select the one that brings the maximum energy savings without performance penalty. It needs to be noted that we may not always select the minimum RPM level for a given array access because there may not be a sufficient number of iterations in the loop nest where this array reference appears.[2] Therefore, at the end of the first step of this approach, the preferred disk speed for each array reference is known. However, if a disk-resident array can be accessed from within multiple loop nests, the disk speed for that array is set to the highest speed among all the preferable speeds for all the references to that array. The algorithm that selects the most suitable disk speeds to be used for each array is given in Figure 2.6. The for-each loop in this algorithm goes over

---

[2] An alternate approach would be inserting the prefetch call for a given loop nest in one of the preceding loop nests, but this makes code generation extremely difficult.

INPUT:
  Input program, $\mathcal{P} = (\mathcal{L}_1, \mathcal{L}_2, \ldots, \mathcal{L}_s)$;
  Available disk speeds, $RPM = (1, 2, \ldots, l)$;
OUTPUT:
  Determined RPM-group(i), where $1 \leq i \leq l$;

T_{pf} = the number of cycles for PF instruction;
for each $V_k \in \mathcal{V}$ // for each array;
  $G[V_k] = \emptyset$; // possible disk speeds for each array;

// repeat for each loop nest $\mathcal{L}_i$.
for each $\mathcal{L}_i \in \mathcal{P}$ {
  s_i = number of cycles need to execute the loop body of $\mathcal{L}_i$;
  for $j = 1$ to l { // for each RPM available
    // repeat for all array reference in $\mathcal{L}_i$
    // assume that $a_i(\vec{I})$ accesses array element $V_k[\vec{F}(\vec{I})]$.
    for each array reference $a_i(\vec{I})$ {
      calculate I/O latency, $T_d(j)$, when RPM is j;
      // determine prefetch distance, d_j, at jth RPM.
      $d_j = \lceil \frac{T_d(j)}{s_i + T_{pf}} \rceil$;
      if (d_j > total number of iterations for the pipeline loop)
        $G[V_k] = G[V_k] \cup \{j\}$;
    }
  }
}

// RPM-group(l) generated by adding maximum value from set $G[V_i]$.
for each array V_i {
  $l = \{x | x \in G[V_i]$ and MAX$(G[V_i])\}$;
  RPM-group(l) = RPM-group(l) $\cup \{V_i\}$;
}

FIGURE 2.6 Disk speed detection algorithm.

the loop nests in the application and the references in them and determines the required disk speed. The for-loop at the end of the algorithm, on the other hand, selects the required RPM level for each array (each V_i). Note that, at the end of this first step, this approach also determines the prefetch distances for all array references in addition to determining the preferable disk speeds for disk-resident arrays as explained previously.

In the next step, the disk layouts of the arrays in the application are determined. To do this, we first form what we call the *RPM groups*. An RPM group holds the arrays that require the same RPM level. Each RPM group is also attached a *weight*, which captures the sum of the number of accesses to the elements of the arrays in that RPM group. The number of disks that will be assigned to each RPM group is then determined by distributing the available disks (actually I/O nodes) across the RPM groups based on their weights in a proportional manner. More specifically, an RPM group with a larger weight is assigned more disks than an RPM group with a lower weight. The reason is that, by assigning more disks to the RPM group with a larger weight, one can exploit the aggregated bandwidth and parallelism presented by multiple disks better. After an RPM group is assigned its disks, the arrays in that group are striped over those disks using conventional striping. The algorithm for determining the disk layouts of arrays is given in Figure 2.7.

The last step of this approach is to restructure the application code to insert prefetch instructions. Since the prefetch distances for all array references have already been determined by the first step explained, the third step uses this information and restructures the application code accordingly based on the strip-mining-based approach proposed by Brown et al. [7]. Figure 2.8 shows the pseudocode for the algorithm that modifies the application code. The overall view of our approach to energy-aware prefetching is depicted in Figure 2.9.

2.2.1.2 Dynamic Compilation Framework

Reference 6 presents an infrastructure that contains a *dynamic optimizing compiler/linker*, a *high-level I/O library* (called HLIOL), a *minidatabase system* (a *metadata manager*), and a *layout manager* that together manage a parallel, hierarchical storage system.[3] The framework provides I/O-optimized access to data sets regardless of the type of media they currently reside on, what their storage layouts are, or where the media are located. Where/how the data sets are stored and in what type of media they are stored are hidden from the user. This allows the user applications to access a data set the same way regardless of its current location and storage layout. The compiler, the

---

[3] An entire hierarchy of a storage system of a parallel machine, which includes local disks with computation nodes, shared disks in the storage system, disk caches, remote disks, as well as the archival tape system as part of the "parallel hierarchical storage system," is considered.

```
INPUT:
    Input program, P = (L₁, L₂, ..., Lₛ);
    Determined RPM-group(i), where 1 ≤ i ≤ l;
OUTPUT:
    Determined data layout for each array;

tot_disks = total number of disks available;
init_disk = 0;
weight[Vᵢ]: the number of accesses made to Vᵢ within P;
weight[V]: the number of accesses made to all arrays within P;

// determine stripe_factor for Vᵢ with same disk speed
// based on the sum of weight[Vᵢ] in RPM-group(i).
for i = 1 to l { // for each RPM-group
    for all Vᵢ ∈ RPM-group(i)
        sum += weight[Vᵢ];
    stripe_factor(Vᵢ) = tot_disks × ⌈ sum/weight[V] ⌉;
    tot_disks −= stripe_factor(Vᵢ);
}

// determine start_disk for each array Vᵢ
// based on the determined stripe_factor for each array.
for i = 1 to l {
    start_disk (Vᵢ) = init_disk;
    init_disk += stripe_factor (Vᵢ);
}
```

FIGURE 2.7   Data layout detection algorithm.

HLIOL, the minidatabase, and the layout manager cooperate to maintain this uniform storage system view.

Figure 2.10 illustrates the major components of the dynamic compilation framework for I/O-intensive parallel applications. The storage system is assumed to be a parallel, hierarchical storage architecture that has typically a disk-based layer such as NAS (network-attached storage), SAN (storage-area network), or may even have an active disk-based system. A tertiary storage (tape system) that serves as the next level in the storage hierarchy is

INPUT:
A loop nest $\mathcal{L}$: for $\vec{I} \in [\vec{L}, \vec{U}]$, step $\vec{b}$ $\langle a_1(\vec{I}), \ldots, a_m(\vec{I}) \rangle$
$\vec{L} = (l_1, l_2, \ldots, l_n)^T$
$\vec{U} = (u_1, u_2, \ldots, u_n)^T$

OUTPUT:
Transformed loop nest $\mathcal{L}'$; for $\vec{I}' \in [\vec{L}', \vec{U}'] \langle a_1(\vec{I}'), \ldots, a_m(\vec{I}') \rangle$

// assume that $\vec{I}_p \in (I_1, I_2, \ldots, I_k)^T$ is the selected pipeline loop
for each I_p selected for V_i {
    add a new controlling loop denoted by II_p $(=[l_p', u_p'])$ to the
        loop nest $\vec{I}$ such that $\vec{I}' = (I_1, \ldots, II_p, I_p, \ldots, I_k)^T$;
    // calculate new loop bounds for II_p and I_p.
    $[l_p', u_p'] = [l_p, u_p]$;
    $b_p' =$ loop step needed to strip-mine I_p loop;
    add b_p' into the loop step vector, $\vec{b}$
        such that $\vec{b}' = (b_1, \ldots, b_p', b_p, \ldots, b_n)$;
    $[l_p, u_p] = [l_p', l_p' + b_p']$;
}
emit "for $\vec{I}' \in [\vec{L}', \vec{U}']$, step $\vec{b}'$ $\langle$";
// insert prefetch instruction.
for all array references being prefetched
    emit "PF($V_i[\vec{F}[\vec{I}']]$)";
// copy loop body from original loop body.
emit "$a_1(\vec{I}'), \ldots, a_m(\vec{I}') \rangle$";
emit "$\rangle$";

FIGURE 2.8 Code restructuring algorithm.

also assumed. In this storage architecture, the most critical issue is to schedule and coordinate accesses to data and manage the data flow between the
different components. It is also assumed that this storage system is used by
parallel applications.

2.2.1.2.1 HLIOL: A High-Level I/O Library

The main goal of this dynamic compilation framework is to identify and implement various I/O optimizations dynamically using the features provided

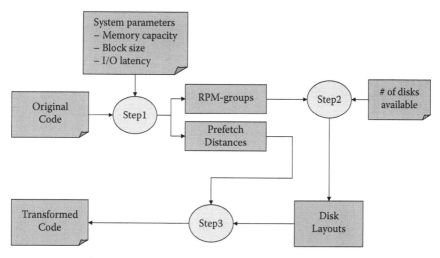

FIGURE 2.9 The three steps in a energy-aware data prefetching compiler framework.

in the HLIOL. The HLIOL's capabilities include an interface that facilities the propagation of I/O access patterns and hints for run-time optimizations. Furthermore, to take advantage of the past access patterns from the application, the HLIOL makes use of a minidatabase (called the metadata manager) that maintains information about the I/O access patterns as well as relationships among data sets. The approach identifies and takes advantage of so-called data set locality, which indicates which data sets tend to be accessed together. The metadata stored in the minidatabase contains such information and is periodically updated during the course of execution. The goal of the minidatabase is to learn and store access patterns at various

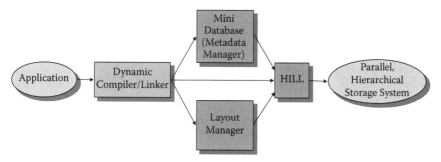

FIGURE 2.10 High-level view of the dynamic compilation approach.

levels and maintain I/O performance statistics. It does not perform I/O. Since the proposed analyses for dynamic compilation are oriented toward exploiting the I/O optimizations supported by the HLIOL, we first explain the HLIOL and briefly discuss its functionality and user interface.

The HLIOL allows an application to access data located in the storage hierarchy via a simple interface expressed in terms of data sets (and arbitrary rectilinear regions of data sets). The main difference between the HLIOL and the previous array-oriented run-time I/O libraries (e.g., Passion [10,11] and Panda [12]) is that the HLIOL maintains the same abstraction (data set name) across an entire storage hierarchy, and that it accommodates storage hierarchy-specific dynamic I/O optimizations.

The routines in the HLIOL can be divided into four major groups based on their functionality: initialization/finalization routines, data access routines, data movement routines, and hint-related routines/queries. Each routine takes a processor ID as one of its input parameters and is invoked by each participating processor. This enables the HLIOL to see the global picture (which includes the I/O access pattern of each processor) in its entirety. Data access routines manage the data flow between storage devices and memory. An arbitrary rectilinear portion of a data set can be read or written using these routines. Using a read routine, for example, the HLIOL can bring a rectangular portion of a data set from tape (or disk) to memory. Data movement routines are used to transfer data between storage devices other than memory. These provide a powerful abstraction by expressing the data movement between any storage device pair as a simple copy operation by working on arbitrary rectilinear portions of data sets. All these routines also have their asynchronous counterparts that return the control to the application code immediately (but perform the specified operation in the background). Queries, on the other hand, are used by the HLIOL to extract specific information from the minidatabase about the data sets, such as their current locations in the storage hierarchy, the sizes of their subfiles, and so on.

The HLIOL contains a large set of I/O optimizations (implemented as library routines) that can be incorporated into the application in an on-demand fashion using dynamic linking. However, if a desired I/O optimization (for the best I/O performance and energy savings) is not available in the HLIOL, the dynamic compiler described in Reference 13 generates the optimized version by making use of the already-available routines (in the HLIOL).

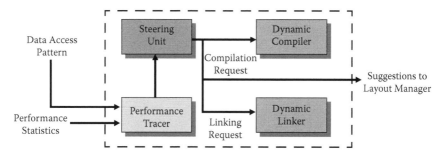

FIGURE 2.11 Components of the dynamic compilation framework.

2.2.1.2.2 Details of the Dynamic Compilation Framework

The dynamic compiler has four major components as depicted in Figure 2.11: (1) dynamic compiler; (2) dynamic linker; (3) performance tracer; and (4) steering unit. The performance tracer is responsible for collecting both I/O access pattern information and performance/energy statistics. The I/O access pattern information includes access directions for data arrays (e.g., row-wise vs. column-wise accesses), whether the data set is accessed in the read-write mode or mostly in the read-only mode, which data sets are accessed with temporal affinity, how frequently the data sets are accessed, and similar information that indicates how different data sets are manipulated by the application. The performance statistics include the number of accesses to different storage units (e.g., tapes, disks), misses in disk/file caches, and the time spent in I/O and the energy consumption in different storage elements.

After collecting this information from the metadata manager, the performance tracer passes it to the steering unit. The main responsibility of the steering unit is to decide whether any dynamic linking or compilation needs to be performed and, if so, select the most appropriate libraries or optimizations to be invoked. As shown in Figure 2.11, the dynamic compiler and linker are invoked by the steering unit.

Table 2.1 lists the I/O optimizations that can be supported by a dynamic compilation framework. The second column briefly describes each optimization, and the third column gives the condition(s) under which each optimization is to be invoked dynamically at run time. We discuss further two of these optimizations in detail, namely, collective I/O and subfiling, and discuss what type of dynamic compiler support needs to be employed for them.

TABLE 2.1   Optimization Rules Incorporated

Optimization	Brief Explanation	Invoked If
Collective I/O (CIO)	Distributing the I/O requests of different processors among them so that each processor accesses as many consecutive data as possible	Access pattern of the data is different from its storage pattern, and multiple processors access the data
Multi-CIO (MCIO)	A version of collective I/O that considers multiple data sets	Data are accessed within the same computation data sets at the same time scope (e.g., a loop nest)
Sequential prefetch (SP)	Bringing consecutive data into higher levels of storage hierarchy before it is needed; overlap I/O time and computation time to hide I/O latency	Access pattern is sequential
Strided prefetch (STD)	Same as sequential prefetching except that data are brought in fixed strides	Access pattern is strided
Replacement policy selection (POL)	Selecting the replacement policy to be used in the higher levels of storage hierarchy (LRU/MRU)	Access size can be captured in cache and is suitable for LRU and MRU (Least Recently Used and Most Recently Used)
Setting striping unit (SSU)	Using a striping unit such that as many storage devices (e.g., disks) as possible will be utilized	Current stripe unit is not performing well
Data migration (DM)	Migrating data from higher levels of storage hierarchy (e.g., disks) to lower levels of storage hierarchy (e.g., tapes)	A particular data set will not be accessed for a long time (but not dead yet)
Data purging (DP)	Removing data from the storage hierarchy; useful for temporal files whose lifetime is over	A particular data set has reached its last use
Prestaging (PRE)	Fetching data from tape subsystem to disk subsystem before it is required	An access to the tape data is predicted
Subfiling (SUB)	Dividing large array into subarrays to reduce transfer latency between different levels of the storage hierarchy	A small subregion of a file is accessed with high temporal locality

In collective I/O, small disk requests are merged into fewer larger requests to minimize the number of times the disks are accessed. While it can be used for both read and write operations, we describe it here only for the read operations. In two-phase I/O [10], a client-side collective I/O implementation, the processors first communicate with each other so that each processor knows the total data that need to be read from the disk system. In the second step, they decide which data each processor needs to read so that the number of disk accesses is minimized. In the next step, the processors perform disk accesses (in parallel). In the last step, they engage in interprocessor communication so that each data item is transferred to its original requester. As an example, consider the access and storage patterns shown in the lower portion of Figure 2.12. Since the access pattern and the storage layout shown are different from each other, allowing each processor to perform its own I/O would lead to numerous I/O requests, each for a small number of array elements. Instead, in two-phase I/O, the processors read the data based on the storage pattern, and this maximizes I/O performance and reduces I/O energy in most cases. After this, they perform interprocessor communication so that the original access pattern is achieved. It needs to be noted that collective I/O, where applicable, can be beneficial from the energy consumption viewpoint since it can reduce the number of disk accesses. While it is true that collective I/O also causes some extra interprocessor data communication, the energy incurred by these communications is normally very small compared to the energy gains achieved on the disk system.

The dynamic compilation analysis for collective I/O has four components: (1) determining I/O access pattern to the data; (2) determining storage pattern (layout) of the data; (3) comparing access and storage patterns to decide whether to apply collective I/O or not; and (4) modifying the code dynamically if necessary. The access pattern information is obtained from the performance tracer, which keeps track of the dynamic I/O access patterns. The storage pattern indicates how the data are stored in the storage system and is maintained by the metadata manager. The steering unit either links the appropriate library routine (in the HLIOL) that implements collective I/O (if such a library routine is available) or dynamically recompiles the application code (that is, the application code is compiled to implement collective I/O using the existing I/O support provided by the HLIOL). This dynamic compilation is confined to the relevant part(s) of the code, that is, typically the loop nest (or a set of related loop nests) that accesses the data in question. Therefore, the energy spent during dynamic compilation is not

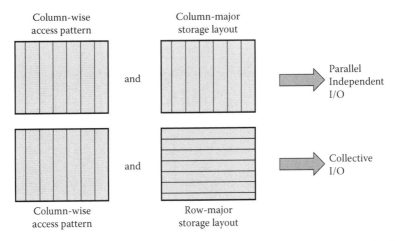

FIGURE 2.12   An example that shows how the steering unit decides whether to apply collective I/O or not.

expected to be excessive. An example of a decision mechanism employed by the steering unit is depicted in Figure 2.12 for two scenarios.

It is also possible that the steering unit may decide a "storage layout (pattern) change" for the data set in question. This may be required in cases where the desired modification to the application code may not preserve the original semantics of the application (hence, it is not legal). In such cases, the steering unit advises the layout manager (see Figures 2.10 and 2.11) to change the storage layout of the data. It should be noted that the layout manager can receive such requests from multiple applications running concurrently on the same storage system, and since a given data set can be accessed by multiple applications, its layout should be modified only if it is going to be beneficial globally (i.e., from multiple applications' perspective). In other words, the steering unit just makes a suggestion (considering only one application), and the layout manager is free to obey it or not.

It should be emphasized that applying I/O optimizations such as collective I/O in a dynamic compilation/linking-based setting brings some unique benefits. For example, in many cases, the data access patterns cannot be extracted statically. Consequently, a static compiler either cannot apply collective I/O (as it does not know the access pattern) or can apply it conservatively, which means reduced energy savings. Also, in some cases, the same data can be shared by multiple applications. It is possible that, between two successive accesses by the same application to the same data set, the layout of data could be modified. In such a case, we need to change the I/O access strategy of the application on the fly to take advantage of the new storage layout. Dynamic compilation and linking allow us to adapt the I/O access behavior to the current status (layout, location) of the data.

The second optimization for which we discuss the necessary dynamic compilation support is subfiling [14]. In many I/O-intensive applications, such as terrain imaging, document imaging, and visualization, although the data sets manipulated are very large, at a given time, only small portions (regions of interest) of the data sets are used. Unfortunately, most current solutions to large-scale data movement across the storage hierarchies proposed by hierarchical storage management systems [15–17] retrieve the entire file that contains the data set in question. This also increases the energy consumption significantly. In fact, this limitation forces the application programmers/users to break their data sets into small, individually addressable objects, thereby cluttering the storage space and making file management very difficult. Instead, subfiling moves the minimum amount of data between storage devices when satisfying a given program's I/O requirements. This is achieved by breaking up the large data sets into uniform, small-size chunks, each of which is stored as a subfile in the storage hierarchy. Therefore, subfiling is expected to bring energy benefits in both tape and disk accesses. Then, an important job of the dynamic compilation framework is to determine the optimal chunk size and restructure the code on the fly based on it. The data access pattern information gives us the type and volume of data reuse. For example, if the accesses are localized (clustered) in small regions of the data set, the chunk size should be kept small; otherwise, we can use a large chunk size. It should also be observed that using subfiling in conjunction with dynamic compilation brings an important advantage over the static compilation-directed subfiling. If we do not use dynamic compilation, then we are forced to select a specific chunk size (most probably based on the profile data), generate code customized for that size, and use that size throughout the execution. In comparison, with the dynamic compilation support, we can change the chunk size during the course of execution, thus better adapting to the dynamic changes in the I/O access patterns.

While dynamic compilation has the potential for improving the performance of I/O-intensive applications and reducing their energy consumptions, it also comes with its own costs that need to be accounted for. Therefore, we should be selective in applying I/O optimizations. However, an overly selective compiler will not work well either as it can miss many optimization opportunities. It is therefore important to maintain cost information within the metadata manager. This cost information consists of the time/energy overhead incurred for each I/O optimization for the last couple of invocations. When the next time the same I/O optimization is needed, the steering unit obtains this cost information from the metadata manager (through the performance tracer) and uses it in deciding whether the optimization in question should really be applied.

2.2.2 Modifying Disk Layout of Data

Reference 18 presents an algorithm for determining energy-aware disk layouts of array data. The goal of this profile-driven algorithm is to increase disk idleness and improve the effectiveness of the underlying disk power management mechanism supported by the hardware.

2.2.2.1 *Abstractions for Disk Layout Optimization*
2.2.2.1.1 Disk Layout Abstraction

File striping is a technique that divides a large amount of data into small portions and stores these portions on separate disks in a round-robin fashion [19]. This permits multiple processes to access different portions of the data concurrently without much disk contention. While striping can be performed manually, many file systems today provide automatic support for it, as discussed in the following material. The disk layout of an array can be represented using a triplet of the form

$$(start\_disk, stripe\_factor, stripe\_size).$$

The first component, start_disk, in this triplet indicates the disk from which the array is started to get striped. The second component, stripe_factor, gives the number of disks used to stripe the data, and the third component, stripe_size, gives the stripe (unit) size used. Several current file systems and I/O libraries for high-performance computing provide application programming interface (APIs) to convey the disk layout information when the file is created.

For example, in PVFS (Parallel Virtual File System) [20], one can change the default striping parameters by setting `base` (the first I/O node to be used), `pcount` (stripe factor), and `ssize` (stripe size) fields of the `pvfs_filestat` structure. Then, the striping information defined by the user via this `pvfs_filestat` structure is passed to the `pvfs_open()` call's parameter. Two example disk layouts for two-dimensional disk-resident arrays are depicted in Figure 2.13. The first layout (i.e., the one for array U) is

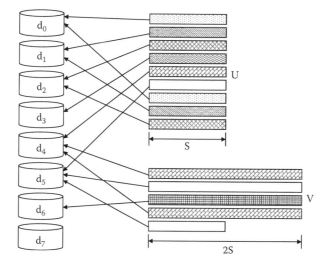

FIGURE 2.13 Two different examples of disk layouts. Left: $(d_0, 6, S)$ and Right: $(d_4, 3, 2S)$.

$(d_0, 6, S)$, whereas the second layout (i.e., the one for array V) is $(d_4, 3, 2S)$. Since a triplet is used for representing disk layout, we can determine the three layout parameters for each disk-resident array that needs to be created by a given application program. It needs to be noted, however that this has to be done in a coordinated fashion by considering all the disk-resident arrays in the application. This is because the different disk-resident arrays can potentially share the same set of disks, and determining their layouts in an independent fashion can lead to unpredictable results (e.g., due to irregular disk access patterns) at run time as far as saving disk power is concerned.

2.2.2.1.2 Power Management Abstraction

Figure 2.14 depicts the transitions between the different states supported by the disks assumed in this study. The labels attached to the arcs in this figure indicate how the transitions are triggered. Each disk is assumed to be equipped with timer-based power management capability. In this mechanism, when the current access to a disk is completed, the disk transitions to the idle state. If it remains in the idle state for a certain amount of time, it is spun down. The disk is said to be placed into the *low-power operation mode*. The disk transitions back to the active mode by spinning up when a new request to it is made. Note that this model represents one of the simplest mechanisms that can be supported by a server disk that allows

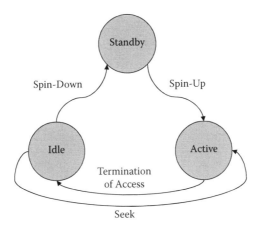

FIGURE 2.14 Different disk states and transitions among them.

power management. While there are other approaches (e.g., [2]), the important point to emphasize here is that, since spinning down and spinning up take both extra time and energy, they need to be minimized. Therefore, one would prefer, from both performance and power consumption angles, a few long idle periods over numerous short idle periods. The goal of the profile-driven approach proposed in Reference 18 is to increase the duration of idle periods, thereby increasing the chances for this power management scheme to be applicable and successful. The proposed approach achieves this by setting the layout parameters for each disk-resident array manipulated by the application code.

2.2.2.2 Disk Layout Detection Algorithm
2.2.2.2.1 High-Level View

Considering energy consumption alone may not be a wise choice since performance of the application is also important and affected significantly by the disk layouts chosen for its arrays. After all, if there was no performance concern, one could potentially work with a single disk, thereby placing all the remaining disks in the system into the low-power operation mode. However, such an approach would hardly be acceptable from the performance perspective. Therefore, the objective is to strike a balance between energy consumption and performance; that is, one would like to save as much energy as possible without significantly impacting original execution cycles, (i.e., execution cycles that would be taken by a pure performance-oriented approach that does not employ any power-saving strategy).

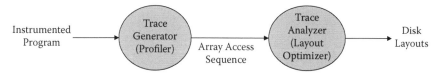

FIGURE 2.15   The connection between profiling and layout optimization.

An important property of the disk layout detection algorithm outlined in Reference 18 is that it is profile driven (as shown in Figure 2.15). The application code is first instrumented and then profiled using a typical input set. The inserted instrumentation code records information on each disk access issued by the application program. At the end of this profiling, an *array access sequence* is obtained, which is of the form $(A_1, A_2, ..., A_n)$. Each array access A_i in this sequence has the form (X, a, t), where X is the ID of the disk-resident array, a is the offset of the accessed array element within the array, and t is the time stamp. The time stamp of an array access is the time since the start of the program after deducting the I/O time spent in the disk accesses. As a simple example, assume a disk access is issued 300 ms after the program starts its execution, and during the first 300 ms of the program execution, disk I/O takes 100 ms total. Then, the time stamp for this array access is calculated as 300 ms − 100 ms = 200 ms. We say that two array accesses $A_i = (X, a_1, t_1)$ and $A_j = (Y, a_2, t_2)$ *conflict* with each other if they access the same disk and the difference between their time stamps is less than the disk response time (denoted by R). Further, the conflicts due to accessing the same array (i.e., $X = Y$) are called the *intraarray conflicts,* and the conflicts due to accessing different arrays (i.e., $X \neq Y$) are referred to as the *interarray conflicts.* Note that, if two array accesses conflict with each other, one of them has to be delayed, which causes the program execution to slow. Based on this discussion, the goal can be rephrased as one of reducing energy consumption on the disk system while minimizing the number of intraarray and interarray conflicts as much as possible.

The approach described in Reference 18 determines the three components of the disk layout of each array in the application: stripe factor, stripe size, and start disk. These three parameters are actually interrelated. That is, electing a value for one of them restricts the potential search space for the other two parameters. Ideally, one would want an algorithm that would try all potential values for all these three parameters of the layout and select the one that generates the best trade-off between energy consumption and performance. However, such an approach is not feasible in general. This

L1: for $i = 0$ to 2047	L2: for $i = 0$ to 1023	L3: for $i = 0$ to 4095
...$X[i]$...;	...$Z[i]$...;	... ...
...$X[i + 1024]$...;	...$Z[i + 256]$...;	...$X[i]$...
...$Y[i]$...;	...$Z[i + 512]$...;	... ...

FIGURE 2.16   An example of code fragment.

is mainly because the search space of potential solutions can be very large. First, in a system with a large number of disks, we have a lot of candidates for the start disk and stripe factor for a given array. Second, one can have many choices for the stripe size, depending on the capabilities of the underlying file system. Third, to reach optimal results, one needs to try all potential layout combinations for all disk-resident arrays. All these factors make an exhaustive search infeasible in practice except for cases with a few disks and a few arrays. Consequently, the profile-driven approach is essentially a fast heuristic that generates not optimal but close-to-optimal results for most cases.

The proposed approach determines a single component of a disk layout at a time. More specifically, it first determines the stripe factor for all arrays and then the stripe size for all arrays. Then, based on these, it determines the start disk for all arrays. The reason that one has to first determine the stripe factor and the stripe size is that these two parameters affect the magnitude of the intraarray conflicts. Once the stripe factor and the stripe size for each array are determined independently, the part of the algorithm that determines the start disk for arrays is executed. This part positions the arrays on the disk system in such a fashion that the number of interarray conflicts is minimized as much as possible.

2.2.2.3   Example

To illustrate how the approach works in practice, we study it by applying it to an example of code fragment. Figure 2.16 presents an example of code fragment. In this example, we set the value of the T parameter to 1, and we use a single file per disk-resident array (1-to-1 mapping). For illustration purposes, let us assume that the underlying I/O system has six disks, the response time for a disk is 5 ms, and each loop iteration (of loops L1, L2, and L3 shown in the figure) takes 10 ms. By analyzing the disk access trace of this program, the approach determines that array X must be stored in at least two disks, and that array Z must be stored in at least three disks. If we stored X in a single disk, the two accesses to X in each iteration of loop L1 would conflict with each other. Specifically, the access to $X[i + 1,024]$ had to wait for the access to $X[i]$ to complete. Similarly, if array Z were stored in fewer than three disks, the three accesses to Z in each iteration of loop L2 would conflict with one another. We can determine the stripe factors for arrays X, Y, and Z as 2, 1, and 3, respectively, since we want to minimize the number of disks used for storing each array without increasing the number of intraarray conflicts (so that energy can be conserved without impacting performance too much).

Stripe size for each array is determined after determining the stripe factor for that array. The number of intraarray conflicts for each array with each possible stripe size is analyzed. Let us assume, again for illustration purposes, that the underlying disk architecture and file system support four stripe sizes: 256B, 512B, 1024B, and 2048B. For array X in loop L1, the number of intraarray conflicts with the stripe sizes 256B, 512B, 1024B, and 2048B, are 2,048, 2,048, 0, and 1,024, respectively. The algorithm selects the stripe size with the minimum number of intraarray conflicts (1024B). The stripe size for array Y does not need to be considered since its stripe factor is one, that is, Y is stored in only a single disk. The number of intraarray conflicts for array Z with stripe sizes 256B, 512B, 1024B, and 2048B are 0, 1,024, 2,048, and 3,072, respectively. Therefore, a stripe size of 256B is selected for array Z.

To determine the start disk for each array, the number of interarray conflicts between each pair of disk stripes is counted. The start disk for each array is then determined, one array after another. In the example code fragment, we have three arrays: X, Y, and Z. We first determine the start disk for array X. This step is trivial since we can pick any disk, say disk 0, as the start disk for array X. And then, we determine the start disk for array Y. At this step, we try all possible start disks for Y and select one that minimizes the interarray conflicts between X and Y. Finally, we determine the start disk for array Z. At this step, we need to select the start disk for array Z such that the total number of interarray conflicts between X and Z and that between Y and Z are minimized. Figure 2.17 gives the final disk layouts determined by this approach for this example, while Figure 2.18 gives another possible disk layout. It is to be emphasized that both these layouts have the same disk conflicts. However, by comparing disk power states[4] presented in Figure 2.19, we observe that the disk layouts determined by this approach exhibit much better idle periods. Specifically, it is seen that this algorithm uses three disks to store all the arrays used in the program so

---

[4] A power-state diagram shows the states of the disks over time.

	Start	Stripe	Stripe
Array	Disk	Factor	Size
X	0	2	1024
Y	2	1	2048
Z	0	3	256

FIGURE 2.17 Disk layout determined by [18].

that the other three disks in the system can remain in the low-power mode throughout the entire execution, and this can lead to significant energy savings at run time. Further, in Figure 2.19, a total of six reactivations is observed (i.e., switching a disk from the low-power mode to the active mode), whereas in Figure 2.19, a total of thirteen reactivations are observed. Note that reactivating a disk incurs both performance and energy penalties.

2.3 ENERGY-AWARE NOC OPTIMIZATIONS

Network-on-chip architectures emerged as an alternate solution to long point-to-point buses in complex multicore processor designs that are going to fuel the future high-performance systems. They have clear advantages over the point-to-point buses from the scalability, maintenance, reliability, and flexibility viewpoints [21–23]. In particular, they address the signal integrity and cost problems associated with long wires. However, prior research [24, 25] showed that NoC power consumption can be a significant portion of the total chip power, making it an important target for both hardware designers and software writers. In particular, scaling voltage/frequency of communication channels in an NoC can lead to quadratic

Array	Start Disk	Stripe Factor	Stripe Size
X	0	4	1024
Y	4	2	1024
Z	0	6	256

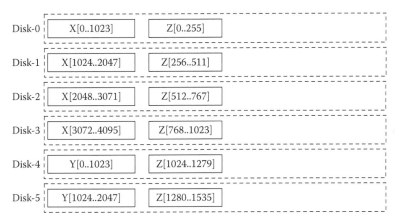

Disk-0 X[0..1023]   Z[0..255]

Disk-1 X[1024..2047]   Z[256..511]

Disk-2 X[2048..3071]   Z[512..767]

Disk-3 X[3072..4095]   Z[768..1023]

Disk-4 Y[0..1023]   Z[1024..1279]

Disk-5 Y[1024..2047]   Z[1280..1535]

FIGURE 2.18 Another possible disk layout.

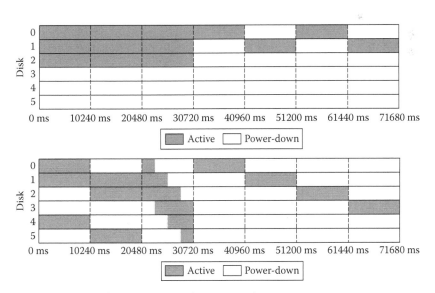

FIGURE 2.19 Disk power states for 2.17 and 2.18.

savings in power but only linear degradation in performance and has been identified as one of the promising mechanisms for saving NoC energy.[5]

Most of the existing power-related studies targeting NoCs are hardware-based efforts. In these studies, a hardware circuit is typically employed to identify channel idleness or variance among channel workloads. The collected information is used for channel shutdown or voltage/frequency scaling. Essentially, similar information on past behavior of NoC message activity can be collected at the granularity of iterations by seeking help from the compiler. While such a compiler-directed technique can reduce power consumption, in a certain class of applications that are amenable to static analysis, one can potentially achieve better results by exploiting compile time information about communication channel utilization and critical execution paths to statically determine channel voltages. In this section, we describe both a history-based compiler scheme and a static communication pattern analysis-based compiler-directed scheme for dynamic link voltage scaling of NoCs to reduce energy consumption.

2.3.1 History-Based Dynamic Link Voltage Scaling

Reference 26 investigated automated compiler support in reducing power consumption of an NoC-based two-dimensional mesh architecture that uses a static (deterministic) routing algorithm. In this approach, NoC is exposed to the compiler through an interface. The goal is to let the compiler modify the application source code and manage power consumption of the communication links through voltage scaling. This approach is based on the observation that the same communication patterns across the nodes of the NoC tend to repeat themselves across the successive iterations of a loop nest. The approach takes advantage of this observation by collecting link usage statistics during the execution of the first few iterations of a given loop nest and computing the allowable delays (slacks) for each communication transaction and the communication bandwidths of the links that are not fully utilized. This information is subsequently utilized in selecting the most appropriate voltage levels for the communication links (and the corresponding frequencies) in executing the remaining iterations of the loop nest [27]. In other words, this compiler-run-time hybrid approach divides the execution of each loop nest that encloses communication among the mesh nodes into two parts. In the first part (called the startup phase), statistics

---

[5]For reasons related to signal integrity and reliability, channel voltage and channel frequency should be scaled together.

```
//On processor 1              //On processor 2
for i = 0 to N {              for i = 0 to N {
    send(2, A [i][0..1024]);      send(1, B [i][0..256]);
    receive(2, buffer);          receive(1, buffer);

    ....                         ....

}                            }
//On processor 1            //On processor 2
for i = 0 to N {            for i = 0 to N {
    send(2, A [i][0..256]);      send(1, B [i][0..256]);
    short computation(...)       long computation(...)
    receive(2, buffer);          receive(1, buffer);


}                            }
```

FIGURE 2.20 Two examples of scenarios.

on link usage at run time are gathered through a hardware-supported interface, and in the second part (called the stable phase), this collected information is used to reduce the link voltage levels as much as possible without affecting communication latency. This approach is similar to run-time-system-based approaches to dynamic voltage scaling [28] but works with the knowledge of the source as the compiler has knowledge of such information. Similar power-aware run-time systems [29–31] utilize DVFS to transparently and automatically adapt CPU voltage and frequency to reduce power consumption in high-performance computing systems.

2.3.2 Static Analysis-Based Link Voltage Scaling

Two example scenarios where a static analysis-based link voltage scheme can be useful are shown in Figure 2.20. In the first scenario in Figure 2.20a, processors in a pair in different NoC nodes communicate with each other using nonblocking send and blocking receive operations. Note that the amounts of data sent between them are different. Consequently, the communication from processor 2 to processor 1 can be performed more slowly than that from processor 1 to processor 2. Therefore, one can potentially scale down the voltage and frequency on the communication channel from processor 2 to processor 1, thereby reducing power. To prevent any performance penalty, the scaling factor should be determined based on the difference between the magnitudes of the communication volumes. In the second scenario in Figure 2.20b, processors in a pair first send data to each other and subsequently perform some computation. Let us assume that, while the data volumes in the two communications are the same, the amount of computation performed by processor 2 is much larger than that performed

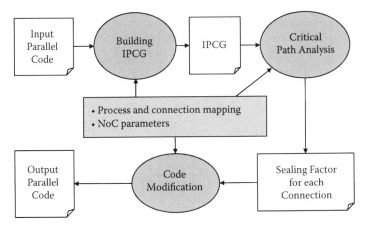

FIGURE 2.21 High-level view.

by processor 1. As a result, one can scale down the voltage/frequency on the communication channel from processor 1 to processor 2. These two scenarios illustrate that the opportunities to scale down voltages can come from the differences between the communication volumes on channels or from the differences between the computation volumes on NoC nodes; both of these variances can be exploited through voltage scaling to reduce NoC power consumption without significantly affecting performance.

Figure 2.21 shows the high-level view of the approach proposed in Reference 32. It assumes that the input application code has already been parallelized (either manually or through a compiler) for message-passing communication, and interprocessor communications have been optimized using known techniques, such as message vectorization and message coalescing [33]. It is also assumed that the process-to-node mapping has already been performed. The input code is first analyzed to build a graph called the *interprocess communication graph* (IPCG), which captures the communication behavior of the parallel application at hand. Then, applying critical path analysis to the IPCG, we can obtain a suitable scaling factor and a voltage level for each communication channel. After that, application code is modified by inserting explicit power management calls.

2.3.2.1 *Interprocess Communication Graph*
An IPCG is a weighted directed graph that is defined for a given message-passing parallel program $\mathcal{P}$ as:

$$G(\mathcal{P}) = (V(\mathcal{P}), E(\mathcal{P}), \alpha, \beta),$$

where $V(\mathcal{P})$ is the set of vertices, $E(\mathcal{P}) \subseteq V(\mathcal{P}) \times V(\mathcal{P})$ is the set of edges, and α and β are the weight functions for the edges. Vertex set $V(\mathcal{P})$ can be expanded as:

$$V(\mathcal{P}) = X(\mathcal{P}) \cup B(\mathcal{P}) \cup S(\mathcal{P}) \cup D(\mathcal{P}) \cup R(\mathcal{P}),$$

where $X(\mathcal{P})$, $B(\mathcal{P})$, $S(\mathcal{P})$, $D(\mathcal{P})$, and $R(\mathcal{P})$ are defined as follows:

$X(\mathcal{P})$ = A vertex $x \in X(\mathcal{P})$ corresponds to a loop in $\mathcal{P}$, and x represents the entry point of this loop.

$B(\mathcal{P})$ = A vertex $b \in B(\mathcal{P})$ corresponds to a loop in $\mathcal{P}$, and b represents the back-jump point of this loop.

$S(\mathcal{P})$ = A vertex $s \in S(\mathcal{P})$ corresponds to a send instruction in $\mathcal{P}$, and s represents the point at which the message is sent.

$D(\mathcal{P})$ = A vertex $d \in D(\mathcal{P})$ corresponds to a send instruction in $\mathcal{P}$, and d represents the point at which the message is delivered to its destination.

$R(\mathcal{P})$ = A vertex $r \in R(\mathcal{P})$ corresponds to a receive instruction in $\mathcal{P}$, and r represents the point at which the message is used by the application.

Note that an interprocess communication in our NoC involves three stages. In the first stage, the sender invokes the send instruction, which copies the message into the buffer of the sender node. The sender process is blocked if this buffer is occupied. Our IPCG captures this using the vertices in $S(\mathcal{P})$. In the second stage, the NoC transfers the message to the receiver (the destination node). This stage completes when all the bits of this message have been delivered to the receiver and stored in a temporary buffer. In the third stage, the receiver invokes a receive instruction to read the contents of the message in the buffer. There may be a gap between the point when a message is delivered to the receiver node and the point when this message is actually accessed by the receiver process. To capture this, our IPCG denotes these two points using different vertices. Specifically, the vertices in $D(\mathcal{P})$ represent the point when a message is delivered to the receiver node, and the vertices in $R(\mathcal{P})$ represent the point when a message is accessed by the receiver process.

We refer to both message send and message receive instructions as *communication instructions*, both communication instructions and back-jumps as *instructions*, and both instructions and loops as *execution units*. We use the term *execution unit* v, where $v \in V(\mathcal{P})$, to refer to the execution unit in program $\mathcal{P}$ that corresponds to vertex v. Further, we use $b(x) \in B(\mathcal{P})$ to denote the back-jump instruction of loop x; $d(s) \in D(\mathcal{P})$, where $s \in S(\mathcal{P})$, to denote the delivery point of sending instruction s; $\psi(v)$, where $v \in V(\mathcal{P})$, to denote the ID of the process to which execution unit v belongs. Further, we write $x \models v$ (where $x \in X(\mathcal{P})$ and $v \in V(\mathcal{P})$) if and only if loop x directly encloses execution unit v. More specifically, if $v \in B \cup S \cup R$, instruction v is in the body of loop x, and it is not enclosed by any other loop nested within loop x. If $v \in X$, loop v is nested within loop x, and there is no other nested loop between loop x and v. In particular, if vertex v is not enclosed by any loop, we write $\phi \models v$. $E(\mathcal{P})$ is the edge set of IPCG $G(\mathcal{P})$. The edges of an IPCG can be classified into seven categories; therefore, we have

$$E(\mathcal{P}) = E_1(\mathcal{P}) \cup E_2(\mathcal{P}) \cup E_3(\mathcal{P}) \cup E_4(\mathcal{P}) \cup E_5(\mathcal{P}) \cup E_6(\mathcal{P}) \cup E_7(\mathcal{P}),$$

where $E_1(\mathcal{P})$ through $E_7(\mathcal{P})$ are defined as follows:

$E_1(\mathcal{P}) = \{(s, d(s)) | s \in S(\mathcal{P})\}$. An edge $(s, d(s)) \in E_1(\mathcal{P})$ is referred to as a *communication edge*.

$E_2(\mathcal{P}) \subseteq X(\mathcal{P}) \times V(\mathcal{P})$. An edge (x, v) is in $E_2(\mathcal{P})$ if $x \models v$ and v is the first execution unit in the body of loop x.

$E_3(\mathcal{P}) \subseteq V(\mathcal{P}) \times V(\mathcal{P})$. An edge (u, v) is in $E_3(\mathcal{P})$ if both u and v are directly enclosed by the same loop x (i.e., $\exists x \in X(\mathcal{P}) : x \models u, v$) and v is executed immediately after u at each iteration of loop x.

$E_4(\mathcal{P}) = \{(b(x), x) | x \in X(\mathcal{P})\}$. An edge $(b(x), x) \in E_4(\mathcal{P})$ is referred to as the *back-jump edge*.

$E_5(\mathcal{P}) \subseteq D(\mathcal{P}) \times S(\mathcal{P})$. An edge $(d(s), s')$ is in $E_5(\mathcal{P})$ if s and s' are directly enclosed by the same loop, and s' is the first send instruction that is executed after s at each loop iteration. Edge $(d(s), s')$ indicates that send instruction s' is blocked by send instruction s, which is executed at same loop iteration. Such an edge is referred to as an *intraiteration blocking edge*.

$E_6(\mathcal{P}) \subseteq D(\mathcal{P}) \times S(\mathcal{P})$. An edge $(d(s), s')$ is in $E_6(\mathcal{P})$ if s and s' are directly enclosed by the same loop, and s and s' are the last and first send instructions in each loop iteration, respectively. Note that, if this loop contains only one send instruction, we have $s = s'$. Edge $(d(s), s')$ indicates that send instruction s' is blocked by send instruction s, which is executed by a previous loop iteration. Such an edge is referred to as an *interiteration blocking edge*.

$E_7(\mathcal{P}) \subseteq D(\mathcal{P}) \times R(\mathcal{P})$. An edge $(d(s), r)$ is in $E_7(\mathcal{P})$ if the messages sent by send instruction s are received by receive instruction r.

Each edge in $E_1(\mathcal{P})$ represents a *communication task*, while each edge in $E_2(\mathcal{P}) \cup E_3(\mathcal{P})$ represents a *computation task*. Therefore, the edges in $E_1(\mathcal{P}) \cup E_2(\mathcal{P}) \cup E_3(\mathcal{P})$ are referred to as the *task edges*. For a task edge (u, v), $\alpha(u, v)$ and $\beta(u, v)$ represent, respectively, the compiler-estimated lower and upper bounds of the length of task (u, v) (i.e., the time it takes to complete this task). The length of a computation task (u, v) is determined by the sum of the latencies of the instructions between the start points of execution units u and v. The values of $\alpha(u, v)$ and $\beta(u, v)$ can be obtained either by profiling or through a static analysis-based approach such as the one proposed in Reference 34. In comparison, the length of a communication task $u, d(u)$ is determined as follows:

$$\alpha(u, d(u)) = \frac{l_{min}}{\lambda} \quad and \quad \beta(u, d(u)) = \frac{l_{max}}{\lambda},$$

where l_{min} and l_{max} are the minimum and maximum sizes of the messages sent by instruction u, respectively; and λ is the maximum available data rate of a communication channel in NoC. On the other hand, the edges in $E_4(\mathcal{P}) \cup E_5(\mathcal{P}) \cup E_6(\mathcal{P}) \cup E_7(\mathcal{P})$ do not represent any real task. They are simply introduced to enforce the timing constraints among instructions; hence, they are referred to as the *control edges*. Consequently, for a control edge (u, v), we have $\alpha(u, v) = \beta(u, v) = 0$.

An edge (u, v) indicates that the execution unit v is executed after execution unit u. Further, an edge $(u, v) \in E_4(\mathcal{P}) \cup E_6(\mathcal{P})$ indicates that u and v are executed in different loop iterations. So, we refer to the edges in set $E_4(\mathcal{P}) \cup E_6(\mathcal{P})$ as the *interiteration edges*. On the other hand, an edge $(u, v) \notin E_4(\mathcal{P}) \cup E_6(\mathcal{P})$ specifies that u and v are executed within the same loop iteration. Therefore, we call the edges not in set $E_4(\mathcal{P}) \cup E_6(\mathcal{P})$ the *intraiteration edges*. Note that the graph obtained by eliminating the interiteration edges from $G(\mathcal{P})$ is acyclic. Therefore, the intraiteration edges in $G(\mathcal{P})$ determine a partial order among the vertices.

2.3.2.2 Determining Scaling Factors and Channel Voltages

2.3.2.2.1 Representative Iterations

For a given parallel group $H = \{x_1, x_2, ..., x_n\}$, let us use $t_{i,j}$ to represent the earliest time that the jth iteration of loop x_i can be started, and we have $t_{i,0} = 0$ ($i = 0, 1, ..., n$). Let us further assume that q is the minimum number such that there exists a constant $R \geq 1$ such that

$$t_{1,q+R} - t_{1,q} = t_{2,q+R} - t_{2,q} = ... = t_{n,q+R} - t_{n,q} = T. \qquad (2.2)$$

The start time of the $(q + mR + k)^{th}$ iteration of loop x_i, where $0 \leq k < R$ and $m \geq 0$, can be computed as $t_{i,q+k} + mT$. Therefore, R can be thought of as the reoccurring period of the timing behavior of parallel group H. The timing behavior of the parallel group H during its entire execution time can be represented by the behavior it exhibits during the period from the qth iteration through the $(q + R - 1)^{th}$ iteration. Therefore, we refer to these iterations as the *representative iterations*. Representative iterations are important because we use them for determining scaling factors.

2.3.2.2.2 Algorithm

Figure 2.22 shows the algorithm for determining the scaling factors for the connections used by a given parallel group H. For an application containing multiple parallel groups, we analyze these parallel groups individually. The algorithm takes $L(H)$, the LCG for the given parallel group H, as input. It computes $k[i, j]$, the scaling factor for the communication channels in the connection from process i to process j, such that the overall percentage performance degradation due to voltage/frequency scaling does not exceed δ, where δ is a user-specified constant, as compared to the performance achieved by operating all the communication channels in NoC at the highest available voltage/frequency level.

In the first phase, the algorithm computes $t_\alpha[i, j]$, the earliest time v_i at the jth iteration can be reached, assuming all the tasks are finished in their shortest times. The values of $t_\alpha[i, j]$s are the minimum values that satisfy the following expressions:

$$\forall i : t_\alpha[i, 0] = 0, \qquad (2.3)$$

$$\forall (k, i) \in E^+ : t_\alpha[i, j] \geq t_\alpha[k, j] + \alpha(k, i), \qquad (2.4)$$

$$\forall (k, i) \in E^- : t_\alpha[i, j] \geq t_\alpha[k, j - 1], \qquad (2.5)$$

Global Variables:
$L(H)$ — the LCG of parallel group H;
V — the set of vertices in LCG $L(H)$. The vertices in V has been sorted in the
      partial order determined by the intra-iteration edges, i.e., for all $v_i, v_j \in V$,
      we have $i < j$ if (v_i, v_j) is an intra-iteration edge.
$t_\alpha[i, j], t_\beta[i, j]$ — the best and the worst start times of vertex v_i in the jth iteration.
$k[i, j]$ — the scaling factor for connection $[[i, j]]$;
      $0 < k[i, j] \le 1$; particularly; $k[i, j] = 1$ if $i = j$.
q, Q, T — $\forall i \in H : T = t[i, Q] - t[i, q]$

// Initialization
set all $t_\alpha[i, j], t_\beta[i, j]$, and $t[i, j]$ to 0;   set all $k[i, j]$ to 1; $Q = 0$; $T = -1$;

// **Phase 1: Computing** $t_\alpha[i, j], q, Q$, **and** T
repeat{
  for $i = 1$ to $|V|$
    for each intra-iteration edge (v_i, v_j) where $t_\alpha[j, Q] < t_\alpha[i, Q] + \alpha(v_i, v_j)$
      $t_\alpha[j, Q] = t_\alpha[i, Q] + \alpha(v_i, v_j)$;
  $Q = Q + 1$;
  for each inter-iteration control edge (v_i, v_j) where $t_\alpha[j, Q] < t_\alpha[i, Q - 1]$
    $t_\alpha[j, Q] = t_\alpha[i, Q - 1]$;
  for $q = Q - 1$ to 0
    if$(\forall v_i, v_j \in H : t_\alpha[i, Q] - t_\alpha[i, q] \ne t_\alpha[j, Q] - t_\alpha[j, q])$ {
      $T = t_\alpha[i, Q] - t_\alpha[i, q]$ where $v_i \in H$;
      break; // the values of T, Q, and q have been determined.
    }
} until $(Q \ge Q^*$ or $T > 0)$;
if$(T = -1)$ {terminate with failure;}

// **Phase 2: Computing** $t_\beta[i, j]$
for $r = q$ to Q { for each $v_i \in V$ { $t_\beta[i, r] = t_\alpha[i, r]$; } }
for $r = q$ to Q {
  for $i = 1$ to $|V|$
    for each intra-iteration edge (v_i, v_j) where $t_\beta[j, r] < t_\beta[i, r] + \beta(v_i, v_j)$
      $t_\beta[j, r] = t_\beta[i, r] + \beta(v_i, v_j)$;
    for each inter-iteration control edge (v_i, v_j) where $t_\beta[j, r + 1] < t_\beta[i, r]$
      $t_\beta[j, r + 1] = t_\beta[i, r]$;
}

// **Phase 3: Computing** $t[i, j]$ **and Determining** $k[i, j]$
for all connection $[[i, j]]$ { $g[i, j] = 0$; }
repeat {
  for each connection $[[i, j]]$ where $g[i, j] = 0$ {
    $k' = k[i, j]$; // back up $k[i, j]$
    decrease $k[i, j]$ to the next scaling level;
    for $r = q$ to Q { for each $v_i \in V$ { $t[i, r] = t_\alpha[i, r]$; } }
    for $r = q$ to Q {
      for $i = 1$ to $|V|$
        for each intra-iteration edge (v_i, v_j)
          where $t[j, r] < t[i, r] + \alpha(v_i, v_j)/k[\psi(v_i), \psi(v_j)]$
            $t[j, r] = t[i, r] + \alpha(v_i, v_j)/k[\psi(v_i), \psi(v_j)]$;
        for each inter-iteration control edge (v_i, v_j) where $t[j, r + 1] < t[i, r]$
          $t[j, r + 1] = t[i, r]$;
    }
    if$(\exists v_i \in H : t[i, Q] - t[i, q] > \max\{(1 + \delta)T, t_\beta[Q, i] - t_\beta[q, i]\})$ {
      $k[i, j] = k'$; // restore $k[i, j]$ to its previous value.
      $g[i, j] = 1$; // this connection cannot be scaled any further.
    }
  }
} until $g[i, j] = 1$ for all connection $[[i, j]]$;

FIGURE 2.22 Algorithm for critical path analysis. This algorithm takes $L(H)$, the LCG of parallel group H as input; it outputs $k[i, j]$, the scaling factor for the communication channels used by each connection $[[i,j]]$.

where E^+ is the set of intraiteration edges, and E^- is the set of interiteration edges. One can observe from the algorithm shown in Figure 2.22 that the first phase contains a repeat-until loop. At the Qth iteration of this loop, we use Expression 2.4 to compute $t_\alpha[i, Q]$ for each vertex v_i in the LCG. After that, we check if there exist a constant q $(0 \leq q < Q)$ and a constant T such that, for all $v_i \in H$, we have $t_\alpha[i, Q] - t_\alpha[i, q] = T$. Note that the qth through $(Q - 1)^{\text{th}}$ iterations are the representative iterations. If we cannot find such a q and T, we use Expression 2.5 to compute the initial values of $t_\alpha[i, Q + 1]$ for the next iteration. We repeat this procedure until we find a suitable q and T. To limit compilation time, the algorithm terminates if it cannot not find suitable q and Q values within the first Q^* iterations. The computational complexity for each iteration of this phase is $O(m + n)$, where m and n are the number of the edges and the vertices in the LCG under consideration, respectively. Therefore, the computational complexity of this phase is $O(Q^*(m + n))$.

In the second phase of the algorithm, the worst-case execution time for the representative iterations is computed assuming that each task takes the longest time to complete and all the communication channels work at the highest data rate. The worst-case start time for each vertex in V is the minimum value of $t_\beta[i, j]$ that satisfies the following expressions:

$$\forall i : t_\beta[i, q] = t_\alpha[i, q], \tag{2.6}$$

$$\forall(k, i) \in E^+ : t_\beta[i, j] \geq t_\beta[k, j] + \beta(k, i), \tag{2.7}$$

$$\forall(k, i) \in E^- : t_\beta[i, j] \geq t_\beta[k, j - 1], \tag{2.8}$$

The worst execution time for the representative iterations of loop $v_i \in H$ can be computed as $t_\beta[i, Q] - t_\beta[i, q]$. The computational complexity for this phase is $O((m + n)(Q - q))$, where m and n are the number of the edges and the vertices in the LCG under consideration, respectively, and the values of q and Q are as determined in the previous phase.

In the third phase, we try to maximize the scaling factor $k[i, j]$ $(0 < k[i, j] \leq 1)$ for each communication channel exercised by connection $[[i,j]]$ under the following constraints:

$$\forall i : t[i, q] = t_\alpha[i, q], \tag{2.9}$$

$$\forall(k, i) \in E^+ : t[i, j] \geq t[k, j] + \beta(k, i)/k[\psi(v_k), \psi(v_i)], \tag{2.10}$$

$$\forall(k, i) \in E^- : t[i, j] \geq t[k, j - 1], \tag{2.11}$$

$$\forall v_i \in H : t[i, Q] \leq t[i, q] + \max\{(1 + \delta)T, t_\beta[i, Q] - t_\beta[i, q]\} \tag{2.12}$$

In this phase, we assume all the tasks are finished in their longest time. Expression 2.12 means that, for a loop whose worst-case overall execution time for the representative iterations does not exceed $(1 + \delta)T$, we can tolerate a percentage performance degradation up to δ. On the other hand, for a loop whose worst-case overall execution time for the representative iterations is already longer than $(1 + \delta)T$, we do not allow any further performance degradation. As a result, the overall performance degradation due to scaling the voltage/frequency of communication channels is within δ.

One can observe from Figure 2.22 that the phase 3 of algorithm contains a "repeat-until" loop. At each iteration of this loop, we select a connection and reduce the data rate of the communication channels in this connection by one level. After that, we estimate the execution time for each loop. If the estimated performance degradation exceeds the limit set by Expression 2.12, the data rate of the selected connection cannot be reduced. We repeat this procedure until there is no connection whose data rate can be further reduced. Note that, at each step, instead of scaling down the voltage/frequency of the selected connection aggressively, we reduce its voltage/frequency by only one level. This approach allows us to scale down the speeds of more connections. The computational complexity of this phase is $O(v^c(m + n)(Q - q))$, where c is the number of connections used, v is the number of available frequency/voltage levels for a communication channel, m is the number of edges, n is the number of the vertices in the LCG under consideration, and the values of q and Q are as determined in phase 1. Therefore, the overall complexity of the algorithm is $O((m + n)(v^c(Q - q) + Q^*))$.

REFERENCES

1. C. Hsu and U. Kremer. The design, implementation, and evaluation of a compiler algorithm for CPU energy reduction. In *Proceedings of the ACM SIGPLAN Conference on Programming Languages Design and Implementation*, 2003.
2. Sudhanva Gurumurthi, Anand Sivasubramaniam, Mahmut Kandemir, and Hubertus Franke. DRPM: Dynamic speed control for power management in server class disks. *SIGARCH Computer Architecture News*, 31(2):169–181, 2003.
3. S. W. Son, M. Kandemir, and A. Choudhary. Software-directed disk power management for scientific applications. In *Proceedings of the 19th International Parallel and Distributed Processing Symposium*, Denver, CO, April 2005.

4. Seung Woo Son and Mahmut Kandemir. Energy-aware data prefetching for multi-speed disks. In *Proceedings of the 3rd Conference on Computing Frontiers*, pages 105–114, Ischia, Italy, May 2006.

5. E. Pinheiro and R. Bianchini. Energy conservation techniques for disk array-based servers. In *Proceedings of the 17th International Conference on Super-computing*, pages 66–78, June 2004.

6. Seung Woo Son, Guangyu Chen, Mahmut Kandemir, and Alok Choudhary. Exposing disk layout to compiler for reducing energy consumption of parallel disk based systems. In *Proceedings of the ACM SIGPLAN Symposium on Principles and Practice of Parallel Programming*, pages 174–185, Chicago, IL, June 2005.

7. A. D. Brown, T. C. Mowry, and O. Krieger. Compiler-based I/O prefetching for out-of-core applications. *ACM Transactions on Computer Systems*, 19(2):111–170, May 2001.

8. M. Wolfe. *High Performance Compilers for Parallel Computing*. Addison-Wesley, Boston, 1996.

9. Michael E. Wolf and Monica S. Lam. A data locality optimizing algorithm. In *Proceedings of the ACM SIGPLAN Conference on Programming Language Design and Implementation*, pages 30–44, Toronto, Canada, June 1991.

10. Alok Choudhary, Rajesh Bordawekar, Michael Harry, Rakesh Krishnaiyer, Ravi Ponnusamy, Tarvinder Singh, and Rajeev Thakur. *Passion: Parallel and Scalable Software for Input-Output*. Technical report, Syracuse University, Syracuse, NY, 1994.

11. Rajeev Thakur, Alok Choudhary, Rajesh Bordawekar, Sachin More, and Sivaramakrishna Kuditipudi. Passion: Optimized I/O for parallel applications. *IEEE Computer*, 29(6):70–78, 1996.

12. K. E. Seamons, Y. Chen, P. Jones, J. Jozwiak, and M. Winslett. Server-directed collective I/O in panda. In *Supercomputing '95: Proceedings of the ACM/IEEE Conference on Supercomputing*, page 57, San Diego, CA, December 1995.

13. Seung Woo Son, Guangyu Chen, Mahmut T. Kandemir, and Alok N. Choudhary. Dynamic compilation for reducing energy consumption of I/O-intensive applications. In *LCPC*, pages 450–457, Hawthorne, NY, October 2005.

14. Gokhan Memik Mahmut, Mahmut T. Kandemir, Alok Choudhary, and Valerie E. Taylor. April: A run-time library for tape-resident data. In *Proceedings of the 17th IEEE Symposium on Mass Storage Systems*, College Park, MD, March 2000.

15. L. T. Chen, R. Drach, M. Keating, S. Louis, D. Rotem, and A. Shoshani. Efficient organization and access of multi-dimensional datasets on tertiary storage systems. *Information Systems Journal*, 20(2):155–183, 1995.

16. Ling Tony Chen, Doron Rotem, Arie Shoshani, Bob Drach, Meridith Keating, and Steve Louis. Optimizing tertiary storage organization and access for spatio-temporal datasets. In *Fourth NASA Goddard Conference on Mass Storage Systems and Technologies*, College Park, pages 165–183, 1995.

17. R. A. Coyne, H. Hulen, and R. Watson. The high performance storage system. In *Supercomputing '93: Proceedings of the ACM/IEEE Conference on Supercomputing*, Portland, OR pages 83–92, November 1993.

18. S. W. Son, G. Chen, and M. Kandemir. Disk layout optimization for reducing energy consumption. In *ICS '05: Proceedings of the 19th Annual International Conference on Supercomputing*, pages 274–283, Cambridge, MA, June 2005.

19. J. M. May. *Parallel I/O for High Performance Computing*. Morgan-Kaufman, Burlington, MA, 2001.

20. R. B. Ross, P. H. Carns, W. B. Ligon III, and R. Latham. Using the Parallel Virtual File System, Technical report, Clemson University, July 2002.

21. L. Benini and G. De Micheli. Networks on chips: A new SoC paradigm. *IEEE Computer*, 35(1):70–78, 2002.

22. William J. Dally and Brian Towles. Route packets, not wires: On-chip inteconnection networks. In *Proceedings of the 38th Conference on Design Automation*, Las Vegas, NV, June 2001.

23. Jingcao Hu and Radu Marculescu. Energy- and performance-aware mapping for regular NOC architectures. *IEEE Transactions on Computer-Aided Design of Integrated Circuits and Systems*, 24(4):551–562, April 2005.

24. V. Soteriou and L.-S. Peh. Design space exploration of power-aware on/off interconnection networks. In *Proceedings of the 22nd International Conference on Computer Design*, Asheville, NC, October 2004.

25. Vassos Soteriou, Noel Eisley, and Li-Shiuan Peh. Software-directed power-aware interconnection networks. In *Proceedings of Conference on Compilers, Architecture and Synthesis for Embedded Systems*, San Francisco, CA, September 2005.

26. Guangyu Chen, Feihui Li, and Mahmut Kandemir. Reducing energy consumption of on-chip networks through a hybrid compiler-runtime approach. In *PACT '07: Proceedings of the 16th International Conference on Parallel Architecture and Compilation Techniques*, Brasov, Romania pages 163–174, September 2007.

27. V. Freeh, N. Kappiah, D. K. Lowenthal, and T. Bletscha. Just-in-time dynamic voltage scaling: Exploiting inter-node slack to save energy in MPI programs. In *Proceedings of the ACM/IEEE SC 2005 Conference*, Seattle, WA November 2003.

28. Sai Prashanth Muralidhara and Mahmut Kandemir. Communication based proactive link power management. In *HiPEAC '09: Proceedings of the 4th International Conference on High Performance Embedded Architectures and Compilers*, Paphos, Cyprus pages 198–215, January 2009.

29. C. Hsu and W. Feng. Effective dynamic voltage scaling through CPU-boundedness detection. In *Lecture Notes in Computer Science*, February 2005.

30. C. Hsu and W. Feng. A feasibility analysis of power awareness in commodity-based high-performance clusters. In *Proceedings of the 7th IEEE International Conference on Cluster Computing (CLUSTER'05)*, Boston, 2005.

31. C. Hsu and W. Feng. A power-aware run-time system for high-performance computing. In *Proceedings of the ACM/IEEE SC*, Seattle, Washington, 2005.

32. Guangyu Chen, Feihui Li, Mahmut Kandemir, and Mary Jane Irwin. Reducing NOC energy consumption through compiler-directed channel voltage scaling. In *PLDI '06: Proceedings of the 2006 ACM SIGPLAN Conference on Programming Language Design and Implementation*, Ottawa, Canada pages 193–203, June 2006.

33. Chau-Wen Tseng. *An optimizing Fortran D compiler for MIMD distributed-memory machines*. PhD thesis, Department of Computer Science, Rice University, Houston, TX, January 1993.

34. Michael E. Wolf, Dror E. Maydan, and Ding-Kai Chen. Combining loop transformations considering caches and scheduling. In *Proceedings of the International Symposium on Microarchitecture*, Paris, France, December 1996.

CHAPTER 3

An Adaptive Run-Time System for Improving Energy Efficiency

Chung-Hsing Hsu, Wu-chun Feng,
and Stephen W. Poole

CONTENTS

3.1 INTRODUCTION

The notion of energy-efficient computing is *not* new, particularly in the areas of embedded systems and mobile computing [1–16], for which computing devices are powered by batteries of limited energy capacity, and reducing energy consumption is critical in extending battery life. Laptops, for example, use simple energy-reduction algorithms to improve their energy efficiency. That is, if a laptop user is reading a document for an extended period of time while running on battery energy, the laptop would automatically scale down the frequency and supply voltage of its CPU (central processing unit) (i.e., processor) to reduce energy consumption. The commodity technology that enables the scaling of frequency and voltage for CPUs is called *dynamic voltage and frequency scaling (DVFS)*. AMD refers to its DVFS mechanism as PowerNow!, while Intel refers to it as SpeedStep.

In contrast, the notion of energy-efficient computing *is* new to the high-performance computing (HPC) community. Why the distinction? First, the computational characteristics found in embedded systems and mobile computing differ markedly from those found in HPC. The use of mobile devices (e.g., laptop) tends to be fairly interactive. As a result, energy-reduction algorithms based on CPU utilization work well on these systems. Unfortunately, with these algorithms laptops fail miserably with respect to HPC applications as the CPU utilization of HPC applications remains very high throughout the entire execution [17]. Second, energy efficiency is needed for different reasons. In embedded and mobile computing, energy efficiency is needed to extend battery life, whereas in HPC, it is needed to reduce the operational costs of powering and cooling HPC systems as well as to reduce their environmental and societal impacts (such as becoming more carbon neutral).

Even more worrisome is the issue of reliability in large-scale HPC systems due to energy inefficiency. Many large-scale HPC systems already consume more power than the cities they are in. Given that the rate at which they are adding computing resources far exceeds the available and planned power capacities, these systems would run out of power capacity very quickly. In addition, as large-scale HPC systems continue to increase in size, the amount of heat generated (and hence, temperature) continues to rise and endangers the system's operation. As a rule of thumb, for every $10°C$ ($18°F$) increase in temperature, the failure rate of a system doubles [18].

There are two main approaches to address the energy efficiency of a HPC system. The first approach relies on the invention of new hardware that can perform the same computation with less energy. The second approach utilizes software-based optimization. Software-based optimization exploits the different levels of impact each software execution pattern has on energy and performance and alters the execution pattern or hardware configuration to achieve energy-efficient computing. Chapter 2 discussed how the software execution pattern can be changed, via code transformations, to reduce energy consumption. In this chapter, we focus on how the hardware configuration can be adapted, through DVFS, to enhance energy efficiency. Chapter 4 will add dynamic concurrency control as another type of hardware configuration to improve energy efficiency.

Specifically, we describe in this chapter an energy-reduction algorithm called the β-adaptation algorithm, its implementation in the run-time system, and its evaluation on commodity HPC platforms, both uniprocessor and multiprocessor. The end result is a run-time system that transparently and automatically adapts CPU voltage and frequency to reduce energy usage while minimizing the impact on performance.

The rest of the chapter is organized as follows: In Section 3.2, we overview different software-based approaches that leverage DVFS. Then in Section 3.3, we present the β-adaptation algorithm, its design and development as a run-time system. Our evaluation strategy for the algorithm and corresponding findings are detailed in Sections 3.4 and 3.5, respectively. Finally, we conclude the chapter in Section 3.6.

3.2   RELATED WORK

Several case studies have demonstrated DVFS as a feasible technology to improve the energy efficiency of HPC systems [19–22]. In these studies, it was shown that many HPC applications have the so-called energy-time trade-off, meaning that a decrease in energy usage is possible, but it comes at the cost of increased execution time. Not every energy-time trade-off is desirable, as some offer little energy savings and large time penalties. However, many applications show a savings that is equal to or better than the penalty (e.g., 20% less energy with 5% more time), and some are much better than that. System software can exploit this energy-time trade-off to simultaneously address the performance requirement as well as create a general energy-aware solution.

With this knowledge that DVFS can indeed be effective in HPC, the next step is to develop various approaches that leverage DVFS. Such approaches include off-line techniques through manual tuning [19–24] and compiler analysis [25–31]. They also include online techniques through dynamic compilation [32–35], MPI (message-passing interface) library-based extensions [36–42], and ubiquitous run-time systems [18, 43–46]. All these approaches determine when to scale down the frequency and supply voltage of the CPUs and do so either

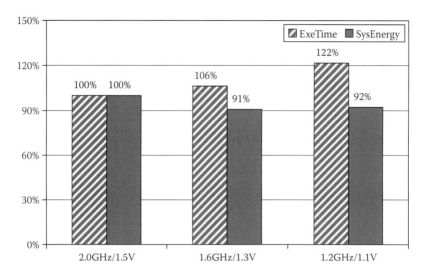

FIGURE 3.1   The performance-power profiles of tomcatv.

off-line or online and target different scenarios, such as during memory access, communication, load imbalance, or their combination.

Manual DVFS tuning often involves profiling of the execution behavior of a program (or its structures) at all possible frequency-voltage settings. It can be as simple as recording the execution time of the program at each available CPU frequency and then using the profile to select the lowest frequency that satisfies the performance constraint to execute the program. Feasibility studies [19–22] fall into this category.

However, this approach is very coarse grained. For example, Figure 3.1 shows the profile for three frequency-voltage combinations on the SPEC (standard performance evaluation cooperation) tomcatv bench-mark. With a 5% performance slowdown constraint in place, the figure indicates that no DVFS setting exists that simultaneously reduces energy consumption and meets the 5% slowdown constraint. The 1.6-GHz/1.3-V and 1.2-GHz/1.1-V settings produce 6% and 22% performance slowdowns, respectively.

A more sophisticated approach is to look into the program structure of the code and profile each interesting program substructure for its execution behavior. For tomcatv, whose program structure is shown in Figure 3.2, this means profiling the execution times of loop nests L1 to L9 at each CPU frequency as tomcatv executes these nested loops in sequence. The tomcatv benchmark spends most of its execution time executing loops L2 to L8 iteratively, with the number of iterations controlled by the variable

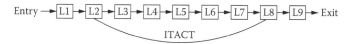

FIGURE 3.2   The program structure of SPEC benchmark `tomcatv`.

ITACT in the code. Freeh et al. [23] used this approach to select the CPU frequency to run for each loop nest and MPI call.

Figure 3.3 shows the execution times of the most time-consuming loops (i.e., L2, L5, L7, and L8) in `tomcatv`. For readability, we normalize all loop execution times with respect to the execution time of the entire benchmark running at 2.0 GHz.[1] The figure indicates that with a 5% performance slowdown constraint, there exist many scheduling options. For example, we can execute loop L2 at 1.6 GHz, resulting in a 3% slowdown, or we can execute loops L5, L7, and L8 at 1.2 GHz.

Although the approach for manual DVFS tuning is straightforward, it can be quite tedious, especially as the number of valid frequency-voltage settings increases, and the program structure becomes more complex. Consequently, automated profiling and subsequent profile analysis are desired. Hsu and Kremer [26] proposed such an implementation based on compiler techniques. In their implementation, compiler techniques such as control-flow-graph analysis are also used to deal with the time overhead caused by setting the CPU frequency and voltage as each such setting takes on the order of milliseconds. For `tomcatv`, their software chooses to slow down loops L6, L7, and L8 to 1.2 GHz. The interested reader can find more details about this off-line compiler-based approach in Reference 47.

The problems with off-line approaches are threefold. First, they are all essentially profile based and generally require the source code to be modified. As a result, these approaches are *not* completely transparent to the end user. Second, because the profile information can be influenced by program input, these approaches are input dependent. Third, as noted in Reference 18, the instrumentation of source code may alter the instruction execution pattern and therefore may produce profiles that are considerably different from the execution behavior of the original code. So, in theory, while these approaches might provide maximal benefit relative to performance and energy, they offer little practical use to end-user applications. As a result, there exists a need for a more transparent and self-adapting run-time approach.

---

[1]That is, at the 2.0-GHz/1.5-V setting, 32% of the execution time is spent in loop L2, 24% in L5, 18% in L7, 18% in L8, and 8% in the remaining loops.

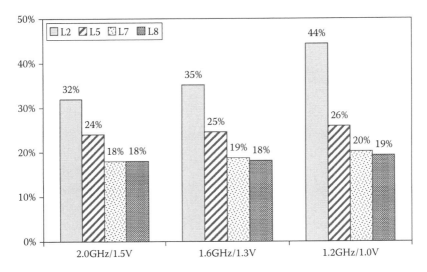

FIGURE 3.3   The execution-time profile of `tomcatv`.

The current approaches toward an adaptive run-time system for energy awareness fall into three categories. The first category is dynamic compilation. A dynamic compiler compiles, modifies, and optimizes a program's instruction sequence as it runs. Thus, dynamic compilation avoids some of the profiling problems off-line approaches have. In addition, similar to off-line approaches, the dynamic compiler adapts naturally to program phase changes. On the other hand, a dynamic compiler consumes energy and affects performance. For most HPC systems, especially those running scientific applications, it is rare to see a dynamic compiler being used.

The second category is MPI library-based extensions. For example, Lim et al. [40] extended the MPI run-time system by intercepting MPI calls to identify communication-bound phases in MPI programs and then scale down the voltage and frequency of the CPUs during these phases. While this approach strives to save energy for a broad class of HPC applications, not all HPC applications are MPI based. Therefore, there exists a need for a ubiquitous run-time approach for all kinds of HPC applications.

The current approaches toward a ubiquitous run-time system for energy awareness are based primarily on CPU utilization (e.g., cpuspeed on laptops). For cpuspeed, when CPU utilization is below some threshold, the CPU voltage and frequency are lowered to conserve energy. When the CPU utilization exceeds some threshold, the CPU voltage and frequency are raised to improve performance. Although this simple approach is both application and input independent, it is only effective for interactive applications (e.g., Microsoft Office) and depends critically on the choice of the threshold values [5]. For scientific applications, its effectiveness is abysmal as such applications do not have an abundance of CPU idle time [17].

Other, more sophisticated, approaches based on CPU utilization, such as those in Reference 48, only provide loose control over DVFS-induced performance slowdown (e.g., 37% slowdown with only 12% system energy savings for the SPEC go benchmark) because the CPU utilization ratio by itself does not provide enough timing information. Therefore, we conclude that there exists a need for an energy-aware run-time system that has tight performance slowdown control and can deliver considerable energy savings.

3.3   AN ADAPTIVE RUN-TIME SYSTEM

Leveraging the DVFS mechanism, we present an automatically adapting, energy-reduction algorithm that is transparent to end-user applications and can deliver considerable energy savings with tight control over DVFS-induced performance slowdown. Performance slowdown in this chapter is defined as the increase in relative execution time with respect to the execution time when the program is running at the peak CPU speed. A user can specify the maximum allowed performance slowdown δ (e.g., $\delta = 5\%$), and our algorithm will schedule CPU frequencies and voltages in such a way that the actual performance slowdown does not exceed δ.

Our energy-reduction algorithm, which we call the β-adaptation algorithm for reasons that become apparent later in the chapter, is an interval-based scheduling algorithm; that is, scheduling decisions are made at the beginning of time intervals of the same length (e.g., every second). Interval-based algorithms are generally easy to implement because they make use of existing "alarm clock" functionality found in the operating system. By default, our energy-reduction algorithm (and its software realization as part of the run-time system) sets the interval length to be 1 s. However, the algorithm allows a user to change this value per program execution. The value is denoted as I hereafter.

In contrast to previous approaches, we want to ensure that our energy-reduction algorithm does not require any application-specific information a priori (e.g., profiling information), and more generally, that it is transparent to end-user applications. Therefore, it must implicitly gather such information, for example, by monitoring the intensity level of off-chip accesses during each interval to make good scheduling decisions. Intuitively, when off-chip access requests are intensive, it indicates that program execution is in a non-CPU-intensive phase, hence implying that this phase can be executed at a lower CPU frequency (and voltage) without affecting program performance.

While conceptually simple, this type of algorithm must overcome the following obstacle to be effective: The quantification of the intensity level of off-chip accesses needs to have a direct correlation between CPU frequency changes and execution-time impact; otherwise, the tight control of DVFS-induced performance slowdown will be difficult to achieve. For example, one might think that the high cache-miss rate is a suitable indicator of program execution in a non-CPU-intensive phase. But, unless we can predict how the execution time will be lengthened for every CPU frequency that may execute this non-CPU-intensive phase, the information of the high cache-miss rate will *not* help selecting an appropriate CPU frequency to maintain tight control of DVFS-induced performance slowdown. *Therefore, we need a model that can associate the intensity level of off-chip accesses with respect to total execution time.*

To overcome the problem, we propose a model that is based on the MIPS (millions of instructions per second) rate, which can correlate the execution-time impact with CPU frequency changes:

$$\frac{T(f)}{T(f_{max})} \approx \frac{\texttt{mips}(f_{max})}{\texttt{mips}(f)} \approx \beta \left(\frac{f_{max}}{f} - 1 \right) + 1. \qquad (3.1)$$

The leftmost term $\frac{T(f)}{T(f_{max})}$ represents the execution-time impact of running at CPU frequency f in terms of the relative execution time with respect to running at the peak CPU frequency f_{max}. The rightmost term $\beta(\frac{f_{max}}{f} - 1) + 1$ introduces a parameter, called β, that quantifies the intensity level of on-chip accesses (and indirectly, off-chip accesses). By definition, $\beta = 1$ indicates that execution time doubles when the CPU speed is halved, whereas $\beta = 0$ means that execution time remains unchanged no matter what CPU speed will be used. Finally, the middle term $\frac{\texttt{mips}(f_{max})}{\texttt{mips}(f)}$ provides a way to describe the observed execution-time impact and is used to adjust the value of β.

Ideally, if we knew the value of β a priori, we could use Equation 3.1 to select an appropriate CPU frequency to execute in the current interval such that the DVFS-induced performance slowdown is tightly constrained. (The selection of this CPU frequency is presented later in the chapter.) But, because we want to ensure that our energy-reduction algorithm does not require any application-specific information a priori, β is *not* known a priori. Therefore, the challenge for our automatically adapting, energy-reduction algorithm lies in the "on-the-fly" estimation of β at run time and hence leads us to name our algorithm the β-adaptation algorithm.

To estimate β at run time, we use a regression method over Equation 3.1 and leverage the fact that most DVFS-enabled microprocessors support a limited set of CPU frequencies to perform the regression. That is, given n CPU frequencies $\{f_1, \cdots, f_n\}$, we derive a particular β value that will minimize the least-squared error:

$$\min \sum_{i=1}^{n} \left\| \frac{\text{mips}(f_{max})}{\text{mips}(f_i)} - \beta \left(\frac{f_{max}}{f_i} - 1 \right) - 1 \right\|^2 \qquad (3.2)$$

By equating the first differential of Equation 3.2 to zero, we can derive β as a function of the MIPS rates and CPU frequencies as follows:

$$\beta = \frac{\sum_{i=1}^{n} \left(\frac{f_{max}}{f_i} - 1 \right) \left(\frac{\text{mips}(f_{max})}{\text{mips}(f_i)} - 1 \right)}{\sum_{i=1}^{n} \left(\frac{f_{max}}{f_i} - 1 \right)^2} \qquad (3.3)$$

Once we calculate the value of β using Equation 3.3, we can plug the value into Equation 3.1 and calculate the lowest CPU frequency f whose predicted performance slowdown $\beta(\frac{f_{max}}{f} - 1)$ does not exceed the maximum possible performance slowdown δ. Mathematically, this establishes the following relationship: $\delta = \beta(\frac{f_{max}}{f} - 1)$. By solving this equation for f, we determine the desired frequency f^* at which the CPU should run:

$$f^* = \max \left(f_{min}, \frac{f_{max}}{1 + \delta/\beta} \right) \qquad (3.4)$$

Combining the aforesaid theory results in the β-adaptation algorithm shown in Figure 3.4. In essence, this energy-reduction algorithm wakes up every I seconds. The algorithm then calculates the value of β using the most up-to-date information on the MIPS rate based on Equation 3.3. Once β is derived, the algorithm computes the CPU frequency f^*, based on Equation 3.4, for the interval. Since a DVFS-enabled microprocessor only supports a limited set of frequencies, the computed frequency f^* may need to be emulated in some cases. (The emulation scheme is shown in Figure 3.5. The ratio r denotes the percentage of time to execute at frequency f_j.) This sequence of steps is repeated at the beginning of each subsequent interval until the program executes to completion.

Hardware:

n frequencies $\{f_1, \cdots, f_n\}$.

Parameters:

I: the time-interval size (default 1 sec).
δ: slowdown constraint (default 5%).

Algorithm:

Initialize $\mathtt{mips}(f_i)$, $i = 1, \cdots, n$, by executing the program at f_i for I seconds.
Repeat

1. Compute coefficient β.

$$\beta = \frac{\sum_i \left(\frac{f_{max}}{f_i} - 1\right)\left(\frac{\mathtt{mips}(f_{max})}{\mathtt{mips}(f_i)} - 1\right)}{\sum_i \left(\frac{f_{max}}{f_i} - 1\right)^2}$$

2. Compute the desired frequency f^*.

$$f^* = \max\left(f_{min}, \frac{f_{max}}{1 + \delta/\beta}\right)$$

3. Execute the current interval at f^*.

4. Update $\mathtt{mips}(f^*)$.

Until the program is completed.

FIGURE 3.4   The β-adaptation algorithm.

To extend the β-adaptation algorithm from the uniprocessor environment that was implicitly assumed previously to a multiprocessor environment, we simply replicate the algorithm onto each processor and run each local copy *asynchronously*. We adopt this strategy for the following reasons: First, the intensity level of off-chip accesses is a per processor metric. Second, a coordination-based, energy-reduction algorithm would need extra communication and, likely, synchronization—both of which add to the time and energy overhead of running the algorithm. As we will see in Section 3.5.2, the β-adaptation algorithm running asynchronously on each processor is quite effective in saving energy while minimizing the impact on performance.

3. Perform the following steps:

   (a) Figure out f_j and f_{j+1}.

$$f_j \leq f^* < f_{j+1}$$

   (b) Compute the ratio r.

$$r = \frac{(1 + \delta/\beta)/f_{max} - 1/f_{j+1}}{1/f_j - 1/f_{j+1}}$$

   (c) Run $r \cdot I$ seconds at frequency f_j.

   (d) Run $(1 - r) \cdot I$ seconds at frequency f_{j+1}.

FIGURE 3.5   Step 3 of β-adaptation algorithm.

In summary, the β-adaptation algorithm is an interval-based, energy-reduction algorithm that is parameterized by two user-tunable variables: the maximum performance slowdown constraint δ and the interval length I, the default values of which are 5% and 1 s, respectively. To facilitate an empirical evaluation of the effectiveness of this algorithm, we implement it in the run-time system, thus creating an energy-aware run-time system. We then test the system on uniprocessor and multiprocessor platforms using appropriate benchmark suites, as discussed in Section 3.4.

3.4   EVALUATION METHODOLOGY AND SETUP

In this section, we describe the evaluation methodology and experimental setup that we used to evaluate the effectiveness of the β-adaptation algorithm.

3.4.1   Evaluation Methodology

To measure the execution time of a program, we use the global time query functions provided by the operating system. In this chapter, the execution time is referred to as the wall clock time of program during execution.

The energy consumption of a program execution is often measured via a power meter. In our experiments, the power meter is connected to a power strip that passes electrical energy from the wall power outlet to the system under test, as shown in Figure 3.6. The power meter periodically samples

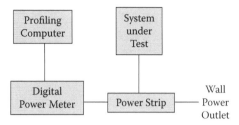

FIGURE 3.6 The experimental setup.

the instantaneous system wattage, and the *total system energy consumption* is then calculated as the integration of these wattages over time. Specifically, we use a Yokogawa WT210 power meter whose sampling rate is 20 μs per sample. The aforementioned integration of wattages is done by the power meter internally.

Unfortunately, evaluating DVFS scheduling algorithms based on total system energy savings can be misleading since DVFS only affects CPU energy consumption. Because the percentage of CPU energy consumption, relative to the total system energy usage, can vary widely from platform to platform, the evaluation results become platform dependent.

For example, consider two DVFS scheduling algorithms, each of which is able to reduce the total system energy by 9%. Intuitively, the two algorithms might be considered equally effective. However, what if one algorithm were evaluated on a HPC server where the CPU accounts for 30% of the total system energy usage while the other algorithm was evaluated on a high-performance laptop computer where the percentage increases to 60%? Back-of-the-envelope calculations[2] show that the former algorithm reduces CPU energy by 30%, and the latter algorithm reduces CPU energy by 15%. Clearly, the former algorithm is more effective than the latter algorithm. This example illustrates that using the platform-dependent, total system energy savings prohibits us from comparing DVFS algorithms evaluated on different platforms fairly. Therefore, we evaluate the effectiveness of the β-adaptation algorithm based on CPU energy savings.

Unfortunately, direct measurements of CPU energy consumption present technical challenges. A common but intrusive method is to place a shunt resistor in series with the CPU and its input power supply. The power meter is connected to this shunt resistor to measure the energy used by the

---

[2]For the first algorithm, the CPU energy savings is calculated as 9%/30% = 30%; for the second algorithm, the savings is 9%/60% = 15%.

CPU [49]. However, intrusive methods based on shunt resistors are argued to be less appropriate because shunt resistors interfere with operation of the system under test and unsuitable when there are large variations in current [50]. As a result, we use a nonintrusive method to estimate the CPU energy consumption.

The nonintrusive method we use leverages a first-order power model for the CPU [51] and divides the system wattage into two parts:

$$P_{sys}(f, V) = \underbrace{C \cdot V^2 \cdot f}_{\text{the CPU power}} + \quad P_{base} \tag{3.5}$$

The first term in the system wattage P_{sys} equation represents the CPU power consumption and depends on the current voltage V and CPU frequency f.[3] The second term P_{base} is independent of voltage and frequency and captures the power consumption of system components that are *not* driven by CPU clock signals.

To estimate the CPU energy consumption for a given application-input pair, we perform a least-squared regression on Equation 3.5 with observation data derived from executing the application-input pair at each possible frequency-voltage combination. This simple approach turns out to be quite accurate when using the R-squared metric. The R-squared metric indicates the relative predictive power of a model; its range is between zero and one, inclusive. The closer the R-squared metric is to one, the more predictive that Equation 3.5 is. For all the benchmarks that we ran in this book chapter, R-squared is very close to one. Therefore, we adopt this unintrusive approach to estimate CPU energy consumption to derive CPU energy savings that the β-adaptation algorithm can deliver.

3.4.2 Systems under Test

In this section, we detail the hardware and software that we used for the performance evaluation of the β-adaptation algorithm in our energy-aware run-time system. We begin by presenting the configurations of the uniprocessor and multiprocessor hardware platforms under test. Then, we describe the systems software on these platforms, followed by information about our implementation of the β-adaptation algorithm. Finally, we list the set of sequential and parallel benchmarks that we used for the evaluation of the algorithm.

---

[3]The constant C in $C \cdot V^2 \cdot f$ denotes the switched capacitance that caused the energy to be consumed. C is application and input dependent.

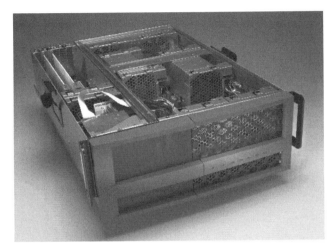

FIGURE 3.7   Celestica A8440.

The tested uniprocessor platform is based on an Asus K8V Deluxe moth-
erboard that is bundled with an AMD Athlon64 3200+ processor (with 1-
MB L2 cache) and 1-GB DDR-400 main memory. The tested multiprocessor
platforms include a cluster of four of the Athlon64-based compute nodes
connected via Gigabit Ethernet and another four-node quad-CPU cluster
based on the Celestica A8440 server (Figure 3.7). server with four AMD
Opteron 846 processors (and also 1-MB L2 cache per processor) and 4-GB
DDR-333 main memory. This Opteron-based cluster is also connected via
Gigabit Ethernet.

In our experiments, both Athlon64 3200+ and Opteron 846 processors
can execute from 800 MHz at 0.9 V to 2 GHz at 1.5 V. Table 3.1 lists the
four operating points (i.e., frequency-voltage pairs) that the β-adaptation
algorithm can set. In theory, an Athlon64 3200+ processor can support
clock frequencies from 800 MHz to 2 GHz at an increment of 200 MHz. So,
why are only four operating points used? It turns out that the set of CPU

TABLE 3.1   Operating Points of
Our Tested Computer Systems

f (GHz)	V
0.8	0.9
1.6	1.3
1.8	1.4
2.0	1.5

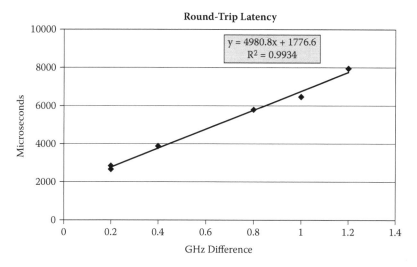

FIGURE 3.8   The latency of each operating point change.

frequencies with direct transitions to each other in an Athlon64 3200+ processor is restricted. Since the time overhead for a direct transition is already on the order of milliseconds (as shown in Figure 3.8), we restrict ourselves to use only a subset of supported CPU frequencies that have direct transitions to each other. For other frequencies, we emulate them using the algorithm in Figure 3.5.

The operating system on the tested hardware platforms is SuSE Linux 2.6.7. This Linux distribution comes with GNU compilers 3.3.3, a DVFS interface called cpufreq, and a DVFS kernel module called powernow-k8. The cpufreq interface allows the β-adaptation algorithm to set the CPU to a desired operating point by writing the target frequency to a particular /sys file. We did not use the powernow-k8 kernel module in the distribution; instead, we used a version of powernow-k8 that is freely downloadable from the AMD website and allows us to specify Table 3.1 in the module.

Our prototype implementation of the β-adaptation algorithm has less than 500 lines of C code. The use of the implementation is similar to the use of a Unix time command. The implementation will fork two threads, one for the execution of the target program (as specified on the command line) and the other for the execution of the β-adaptation algorithm. Thus, our performance evaluation includes the time and energy overhead of running the β-adaptation algorithm in addition to the normal program execution. In practice, the overhead is negligible as the algorithm is lightweighted.

With respect to the benchmarks, we used the SPEC CFP95 and CPU2000 benchmarks for the uniprocessor platform and the NAS-MPI benchmarks, version 3.2, for the multiprocessor platforms. With the exception of SPEC CPU2000 benchmarks, all the other benchmarks were compiled using the GNU compiler 3.3.3 with optimization level -O3. The CPU2000 benchmarks were compiled using the Intel compiler 8.1 with the optimization level -xW -ip -O3. We used the Intel compiler, instead of the GNU compiler, because CPU2000 contains several FORTRAN-90 codes that the GNU 3.3.3 compiler does not support. For the MPI benchmarks, LAM/MPI version 7.0.6 was used to run the benchmarks.

3.5   EXPERIMENTAL RESULTS

This section presents a performance evaluation of the β-adaptation algorithm as a run-time system. As noted, we evaluated the algorithm in both uniprocessor and multiprocessor environments. We also compared the performance of the algorithm against a compiler-based approach.

3.5.1   Uniprocessor Platform

In this section, we present experimental results of our energy-aware run-time system on an Athlon64-based computer using the SPEC CPU200 benchmark suite. Figure 3.9 shows the actual performance slowdown and the CPU energy savings delivered by the system. The β-adaptation algorithm reduces the CPU energy consumption by 12% (on average) with only 4% actual performance slowdown for SPEC CFP2000. For SPEC CINT2000, the two numbers are 9.5% and 4.8%, respectively. Given that the average β values for CFP2000 and CINT2000 are 0.66 and 0.83, respectively, CINT2000 is more CPU bound than CFP2000 and therefore has fewer opportunities for energy savings due to the correspondingly fewer off-chip accesses.

3.5.2   Cluster Platform

In this section, we present experimental results on two clusters: an Athlon64-based cluster and an Opteron-based cluster.

For the Athlon64-based cluster, Figure 3.10a shows the average β value for each of the eight NAS-MPI benchmarks as well as the associated R-squared metric for the class B workload. (Recall that the larger the β, the more CPU bound the benchmark is.) The β value of the benchmarks spans

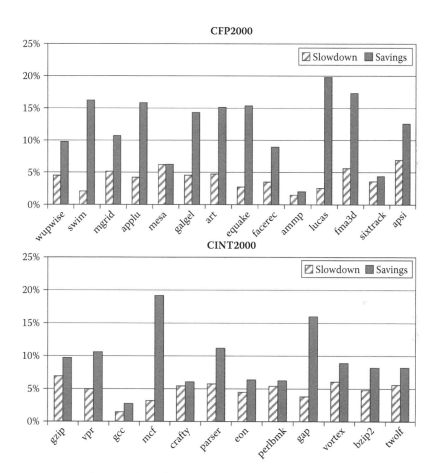

FIGURE 3.9 The actual performance slowdown and CPU energy savings of CPU2000 benchmarks using our run-time system.

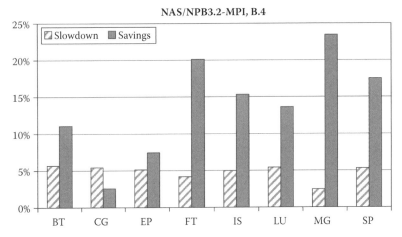

FIGURE 3.10 NAS-MPI for class B workload on the Athlon64-based cluster.

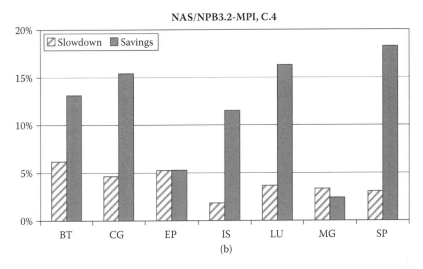

FIGURE 3.11   NAS-MPI for class C workload on the Athlon64-based cluster.

from 0.33 (IS benchmark) to 1.00 (CG and EP benchmarks) with an average value around 0.57. Compared to the SPEC CPU2000 benchmarks, the NAS-MPI benchmarks are generally less CPU bound, which means more opportunities that can be exploited by the run-time system for CPU energy reduction under the same performance slowdown constraint δ.

Figure 3.10b shows the actual performance slowdown and CPU energy savings of NAS-MPI for the class B workload. On average, the β-adaptation algorithm saves 14% CPU energy at 5% performance slowdown. For the class C workload (with larger problem sizes than the class B workload), the average savings is about 12% at 4% slowdown, as shown in Figure 3.11. Note that the result of the FT benchmark is not presented here. This is because the execution of the benchmark had unusually high I/O (input/output) activities due to the lack of memory, which we do not feel is representative for typical HPC applications.

For the Opteron-based cluster, Figure 3.12 shows that the β-adaptation algorithm was able to save CPU energy ranging from 8% to 25%, with an average savings of 18%. The average performance slowdown was 3%. Note that for the MG benchmark, the performance is in fact improved when running at a lower clock speed.

Finally, we want to make a few comments regarding the opportunities that DVFS can exploit. First, a CPU-bound benchmark is not necessarily more power hungry. For example, Figure 3.13 shows the CPU power consumption of NAS-MPI for both class B and C workloads. According to

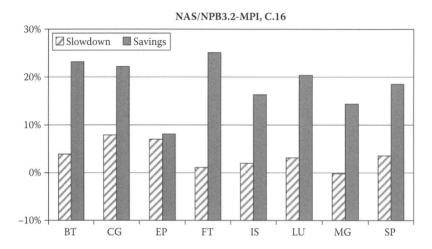

FIGURE 3.12   NAS-MPI for class C workload on the Opteron-based cluster.

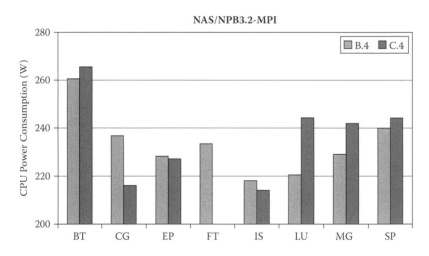

FIGURE 3.13   CPU power consumption of NAS-MPI.

Figure 3.10, EP is a CPU-bound benchmark. However, Figure 3.13 shows that EP is comparably less power hungry than other NAS-MPI benchmarks. The most power-hungry NAS-MPI benchmark for class B workload is BT.

Second, MPI point-to-point communication operations may not be good places to lower the CPU frequency. Given that the latency for setting to another operation point is 1.67 ms or more, it is on the same order of magnitude as the latency for a single MPI point-to-point operation. On our Opteron-based cluster, the typical latency for a single MPI point-to-point operation is less than 1.25 ms for message sizes no greater than 64 KB on a gigabit network. This latency will be shorter on faster networks such as InfiniBand, Myrinet, and Quadrics.

3.5.3 Run-Time versus Compiler-Based Approach

In this section, we compare the performance of the β-adaptation algorithm (running in our energy-aware run-time system) with the compiler-based approach presented in Reference 26 for the SPEC CFP95 benchmarks. Although the CFP95 benchmarks have been retired, they allow us to compare the results of the β-adaptation algorithm to previous case studies [17, 26].

Figure 3.14 shows a comparison of the actual performance slowdown between the run-time approach (denoted as beta) and the compiler-based approach (denoted as hsu) for the maximum performance slowdown of 5%. Here, we see that the actual performance slowdown induced by the

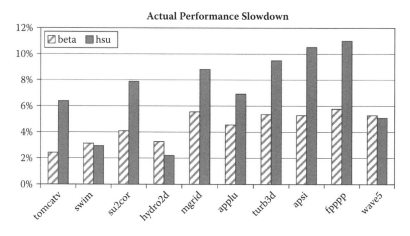

FIGURE 3.14 The actual performance slowdown of the β-adaptive run-time approach versus a compiler-based approach.

compiler approach is poorly regulated. In contrast, the β-adaptation algorithm, which is the foundation of our energy-aware run-time system, regulates the actual performance slowdown much better.

Further investigation reveals that the benchmarks that cause the compiler approach to induce unacceptable performance slowdown (i.e., `mgrid`, `turb3d`, and `apsi`) have CPU-bound execution behavior. This implies that the β-adaptation algorithm for our system will perform more effectively on CPU-bound programs than the compiler approach. Empirical results from a laptop computer [17] corroborated that conclusion.

We believe that the effectiveness of our run-time approach is due to the validity of Equation (3.1). If we apply the least-squared regression on the equation using the *overall* execution time at various CPU frequencies for CFP95, we will see that R-squared is close to one for every CFP95 benchmark. In other words, Equation (3.1) provides a good model for relating the execution time impact with CPU frequency changes.

On the other hand, the ineffectiveness of the compiler approach may be due to the skewed profile information. When the CFP95 benchmarks were run on our HPC servers, the intensity of off-chip accesses in each program construct dropped significantly due to the larger caches in server processors. As a result, program constructs became less distinguishable by the compiler approach, which magnified the instrumentation-altered performance skew and made the compiler difficult to find a definite "winner."

Relative to CPU energy reduction, previous studies (e.g., [17, 26]) reported an average CPU energy reduction of 20% using the compiler approach on a laptop computer. In contrast, the average CPU energy savings for our uniprocessor HPC server platform using our automatically adapting software is about 11%. The gap between the two energy-saving values is due to the difference in L2 cache size. For the laptop, the L2 cache size is only 256 KB, whereas the L2 cache size for the HPC server is four times larger at 1 MB. Consequently, the intensity of off-chip accesses for the laptop is significantly higher than for the HPC server, thus providing substantially more opportunities for energy savings for the laptop.

3.6   CONCLUSION

Power awareness has increasingly become an important issue in HPC. In HPC, ignoring power consumption as a design constraint results in a system with high operational costs for power and cooling and can detrimentally impact reliability, which translates into lost productivity.

To address these issues, we described an energy-aware solution that works on any commodity platform that supports DVFS. Specifically, we presented an energy-reduction algorithm called the β-adaptation algorithm and a prototype implementation of the algorithm as an energy-aware run-time system. The system transparently and automatically adapts CPU voltage and frequency to reduce power consumption (and energy usage) while minimizing impact on performance. The performance evaluation on both uniprocessor and multiprocessor platforms showed that the system achieves its design goal. That is, the system can save CPU energy consumption by as much as 20% for sequential benchmarks and 25% for parallel benchmarks that we tested, at a cost of 3–5% performance degradation. Moreover, the performance degradation was tightly controlled by our run-time system for all the benchmarks.

ACKNOWLEDGMENTS

This work was partially supported by the Extreme Scale Systems Center at Oak Ridge National Laboratory. The submitted manuscript was authored by a contractor of the U.S. government under contract no. DE-AC05-00OR22725. Accordingly, the U.S. government retains a nonexclusive, royalty-free license to publish or reproduce the published form of this contribution, or allow others to do so, for U.S. government purposes.

REFERENCES

1. K. Flautner, S. Reinhardt, and T. Mudge. Automatic performance-setting for dynamic voltage scaling. In *International Conference on Mobile Computing and Networking*, July 2001.
2. J. Flinn and M. Satyabarayanan. Energy-aware adaptation for mobile applications. In *ACM Symposium on Operating Systems Principles*, December 1999.
3. K. Govil, E. Chan, and H. Wasserman. Comparing algorithms for dynamic speed-setting of a low-power CPU. In *International Conference on Mobile Computing and Networking*, November 1995.
4. F. Gruian. Hard real-time scheduling for low-energy using stochastic data and DVS processors. In *International Symposium on Low Power Electronics and Design*, August 2001.
5. D. Grunwald, P. Levis, K. Farkas, C. Morrey III, and M. Neufeld. Policies for dynamic clock scheduling. In *USENIX Symposium on Operating System Design and Implementation*, October 2000.

6. J. Lorch and A. Smith. Improving dynamic voltage algorithms with PACE. In *International Joint Conference on Measurement and Modeling of Computer Systems*, June 2001.

7. B. Mochocki, X. Hu, and G. Quan. A unified approach to variable voltage scheduling for nonideal DVS processors. *IEEE Transactions on Computer-Aided Design of Integrated Circuits and Systems*, 23(9):1370–1377, September 2004.

8. T. Pering, T. Burd, and R. Brodersen. Voltage scheduling in the lpARM microprocessor system. In *International Symposium on Low Power Electronics and Design*, July 2000.

9. N. Pettis, L. Cai, and Y.-H. Lu. Dynamic power management for streaming data. In *International Symposium on Low Power Electronics and Design*, August 2004.

10. P. Pillai and K. Shin. Real-time dynamic voltage scaling for low-power embedded operating systems. In *USENIX Symposium on Operating System Design and Implementation*, October 2001.

11. J. Pouwelse, K. Langendoen, and H. Sips. Dynamic voltage scaling on a low-power microprocessor. In *International Conference on Mobile Computing and Networking*, July 2001.

12. Y. Shin, K. Choi, and T. Sakurai. Power optimization of real-time embedded systems on variable speed processors. In *International Conference on Computer-Aided Design*, November 2000.

13. T. Simunic, L. Benini, and G. De Micheli. Dynamic power management of portable systems. In *International Conference on Mobile Computing and Networking*, August 2000.

14. M. Weiser, B. Welch, A. Demers, and S. Shenker. Scheduling for reduced CPU energy. In *USENIX Symposium on Operating System Design and Implementation*, November 1994.

15. R. Xu, C. Xi, R. Melhem, and D. Mossé. Practical PACE for embedded systems. In *International Conference on Embedded Software*, September 2004.

16. H. Zeng, X. Fan, C. Ellis, A. Lebeck, and A. Vahdat. ECOSystem: Managing energy as a first class operating system resource. In *International Conference on Architectural Support for Programming Languages and Operating Systems*, October 2002.

17. C. Hsu and W. Feng. Effective dynamic voltage scaling through CPU-boundedness detection. In *Workshop on Power-Aware Computer Systems*, December 2004.

18. C. Hsu and W. Feng. A power-aware run-time system for high-performance computing. In *International Conference for High Performance Computing, Networking, Storage and Analysis*, November 2005.

19. X. Feng, R. Ge, and K.W. Cameron. Power and energy profiling of scientific applications on distributed systems. In *International Symposium on Parallel and Distributed Processing*, April 2005.

20. V.W. Freeh, D.K. Lowenthal, F. Pan, N. Kappiah, and R. Springer. Exploring the energy-time tradeoff in MPI programs on a power-scalable cluster. In *International Symposium on Parallel and Distributed Processing*, April 2005.

21. C. Hsu and W. Feng. A feasibility analysis of power awareness in commodity-based high-performance clusters. In *IEEE International Conference on Cluster Computing*, September 2005.

22. F. Pan, V.W. Freeh, and D.M. Smith. Exploring the energy-time trade-off in high-performance computing. In *Workshop on High-Performance, Power-Aware Computing*, April 2005.

23. V.W. Freeh, D.K. Lowenthal, F. Pan, and N. Kappiah. Using multiple energy gears in MPI programs on a power-scalable cluster. In *ACM SIG-PLAN Symposium on Principles and Practice of Parallel Programming*, June 2005.

24. R. Ge, X. Feng, and K.W. Cameron. Improvement of power-performance efficiency for high-end computing. In *Workshop on High-Performance, Power-Aware Computing*, April 2005.

25. M.A. Ghodrat and T. Givargis. Efficient dynamic voltage/frequency scaling through algorithmic loop transformation. In *International Conference on Hardware-Software Codesign and System Synthesis*, October 2009.

26. C. Hsu and U. Kremer. The design, implementation, and evaluation of a compiler algorithm for CPU energy reduction. In *ACM SIGPLAN Conference on Programming Languages Design and Implementation*, June 2003.

27. A. Rangasamy, R. Nagpal, and Y.N. Srikant. Compiler-directed frequency and voltage scaling for a multiple clock domain microarchitecture. In *International Conference on Computing Frontiers*, May 2008.

28. J. Shirako, M. Yoshida, N. Oshiyama, Y. Wada, H. Nakano, H. Shikano, K. Kimura, and H. Kasahara. Performance evaluation of compiler controlled power saving scheme. In *International Workshop on Advanced Low Power Systems*, September 2005.

29. K. Shyam and R. Govindarajan. Compiler-directed dynamic voltage scaling using program phases. In *International Conference on High Performance Computing*, December 2007.

30. F. Xie, M. Martonosi, and S. Malik. Compile time dynamic voltage scaling settings: Opportunities and limits. In *ACM SIGPLAN Conference on Programming Languages Design and Implementation*, June 2003.

31. H. Yi, J. Chen, and X. Yang. Compiler-directed energy-time tradeoff in MPI programs on DVS-enabled parallel systems. In *International Symposium on Parallel and Distributed Processing and Applications*, December 2006.

32. Q. Shi, T. Chen, X. Liang, and J. Huang. Dynamic compilation framework with DVS for reducing energy consumption in embedded processors. In *International Conference on Embedded Software and Systems*, August 2008.

33. S.W. Son, G. Chen, M. Kandemir, and A. Choudhary. Dynamic compilation for reducing energy consumption of I/O-intensive applications. In *International Workshop on Languages and Compilers for Parallel Computing*, October 2005.

34. Q. Wu, M. Martonosi, D.W. Clark, V.J. Reddit, D. Connors, Y. Wu, J. Lee, and D. Brooks. A dynamic compilation framework for controlling microprocessor energy and performance. In *International Symposium on Microarchitecture*, November 2005.

35. L. Xiang, J. Huang, W. Sheng, and T. Chen. The design and implementation of the DVS based dynamic compiler for power reduction. In *International Symposium on Advanced Parallel Processing Technologies*, November 2007.

36. M. Etinski, J. Corbalan, J. Labarta, M. Valero, and A. Veidenbaum. Power-aware load balancing of large scale MPI applications. In *Workshop on High-Performance, Power-Aware Computing*, May 2009.

37. V.W. Freeh, N. Kappiah, D.K. Lowenthal, and T.K. Bletsch. Just-in-time dynamic voltage scaling: Exploiting inter-node slack to save energy in MPI programs. *Journal of Parallel and Distributed Computing*, 68(9):1175–1185, September 2008.

38. Y. Hotta, M. Sato, H. Kimura, S. Matsuoka, T. Boku, and D. Takahashi. Profile-based optimization of power performance by using dynamic voltage scaling on a PC cluster. In *Workshop on High-Performance, Power-Aware Computing*, April 2006.

39. N. Kappiah, V.W. Freeh, and D.K. Lowenthal. Just in time dynamic voltage scaling: Exploiting inter-node slack to save energy in MPI programs. In *International Conference for High Performance Computing, Networking, Storage and Analysis*, November 2005.

40. M.Y. Lim, V.W. Freeh, and D.K. Lowenthal. Adaptive, transparent frequency and voltage scaling of communication phases in MPI programs. In *International Conference for High Performance Computing, Networking, Storage and Analysis*, November 2006.

41. B. Rountree, D.K. Lowenthal, B.R. de Supinski, M. Schulz, V.W. Freeh, and T. Bletsch. Adagio: making DVS practical for complex HPC applications. In *International Conference on Supercomputing*, June 2009.

42. W. Yang and C. Yang. Exploiting energy saving opportunity of barrier operation in MPI programs. In *Asia International Conference on Modelling and Simulation*, May 2008.

43. R. Ge, X. Feng, W. Feng, and K.W. Cameron. CPU MISER: A performance-directed, run-time system for power-aware clusters. In *International Conference on Parallel Processing*, September 2007.

44. S. Huang and W. Feng. Energy-efficient cluster computing via accurate workload characterization. In *International Symposium on Cluster Computing and the Grid*, May 2009.

45. K. Malkowski, G. Link, P. Raghavan, and M.J. Irwin. Load miss prediction—exploiting power performance trade-offs. In *Workshop on High-Performance, Power-Aware Computing*, March 2007.

46. K. Malkowski, P. Raghavan, M. Kandemir, and M.J. Irwin. Phase-aware adaptive hardware selection for power-efficient scientific computations. In *International Symposium on Low Power Electronics and Design*, August 2007.

47. C. Hsu and U. Kremer. Compiler support for dynamic frequency and voltage scaling. In J. Henkel and S. Parameswaran, editors, *Designing Embedded Processors: A Low Power Perspective*. Kluwer Academic Press, Dordrecht, Netherland 2007.

48. A. Varma, B. Ganesh, M. Sen, S. Choudhary, L. Srinivasan, and B. Jacob. A control-theoretic approach to dynamic voltage scaling. In *International Conference on Compilers, Architectures, and Synthesis for Embedded Systems*, October 2003.

49. J. Seng and D. Tullsen. The effect of compiler optimizations on Pentium 4 power consumption. In *Workshop on Interaction between Compilers and Computer Architectures*, February 2003.

50. A. Milenkovic, M. Milenkovic, E. Jovanov, and D. Hite. An environment for runtime power monitoring of wireless sensor network platforms. In *IEEE Southeastern Symposium on System Theory*, March 2005.

51. T. Mudge. Power: A first class design constraint for future architectures. *IEEE Computer*, 34(4):52–58, April 2001.

CHAPTER 4

Energy-Efficient Multithreading through Run-Time Adaptation

Matthew Curtis-Maury and Dimitrious S. Nikolopoulus

CONTENTS

4.1   INTRODUCTION

Processor vendors have given up power-inefficient single-core designs in favor of designs with many simple and power-efficient cores. Parallelism can improve power efficiency and assist the system designer in trading off power with performance. The conventional wisdom holds that with more concurrency, performance improves, but the processor consumes more power. Conversely, with less concurrency, performance drops, but the processor consumes less power.

There are certain cases where inherent workload characteristics (e.g., limited algorithmic concurrency, fine-grain parallel tasks, and synchronization) and architectural properties (e.g., capacity limitations of shared resources) limit the scalability and the maximum degree of exploitable concurrency in an application, resulting in an observed performance *loss* through the use of *more* concurrency. Therefore, throttling concurrency has the potential to both *improve performance* and *reduce power consumption* simultaneously.

Dynamic concurrency throttling (DCT) and dynamic voltage and frequency scaling (DVFS) are two software-controlled mechanisms, or *knobs*, for run-time power-performance adaptation on systems with multicore processors. Earlier research, including our own significantly improved understanding of performance and power implications of DVFS [15, 23] and DCT [4, 5] in isolation. This chapter reviews the aforementioned research and explores methods to synthesize DVFS and DCT techniques in an integrated power-performance adaptation framework. In particular, this chapter explores methods to integrate power-performance adaptation in thread-based parallel programming models.

Synthesizing two power-performance adaptation knobs in software is nontrivial because the search space for adaptation can grow to unmanageable proportions. An M-core processor with L voltage/frequency levels presents a power-performance adaptation search space of size $O(L \cdot M)$. If we assume that processor cores are heterogeneous or cores are homogeneous but performance is sensitive to the placement of threads to cores (this is the case in NUMA (nonuniform memory access) and NUCA (nonuniform cache access) systems, for example), the search space grows to $O(L \cdot 2^M)$. Furthermore, if we assume that a program executes through multiple phases with distinct optimal concurrency and placement of threads to cores, the search space grows linearly with the number of phases in the program, and complex interactions between program phases need to be taken into account. Software cannot reasonably search this space efficiently at run time, even on processors with a modest number of cores and power states. The problem becomes even more pronounced in the context of emerging many-core architectures, such as Intel's 80-core prototype [32] and 48-core Single-Chip Cloud Computer [13].

We present a software framework for multidimensional, software-controlled and performance-constrained adaptation of OpenMP (Open Multi-Processing) programs on multicore systems. Its key component, a dynamic multidimensional performance predictor, statistically analyzes samples of hardware event rates collected from performance monitors and predicts the performance impact of setting DCT and DVFS levels to any of the values available on the system, either combined or in isolation. The statistical analysis uses a rigorous regression model, trained from samples of the power-performance adaptation search space collected from real workloads. The model has low cost, is independent of the application to which it is applied, and is independent of input. These properties enable the use of the model at run time as a dynamic optimization tool. We demonstrate these capabilities through experiments, showing that the presented framework enables reduction of power consumption, energy, and execution time simultaneously, while outperforming both static and dynamic search-based power-performance adaptation frameworks.

The rest of this chapter is organized as follows: Section 4.2 discusses background and related work on power-performance adaptation in run-time systems. Section 4.3 introduces and evaluates the proposed dynamic, phase-aware performance predictor. Section 4.4 presents an implementation of multidimensional (DCT, DVFS) power-performance adaptation that uses the performance predictor. Section 4.5 presents our conclusions.

4.2 RUN-TIME SYSTEMS FOR POWER-PERFORMANCE ADAPTATION: AN OVERVIEW

4.2.1 Dynamic Voltage and Frequency Scaling

Most run-time systems for energy-aware program execution use DVFS to reduce processor power consumption while simultaneously attempting to control losses in performance. Algorithms for DVFS exploit processor idle time using three approaches. The first quantifies memory boundedness to apply DVFS during execution phases with high memory latency [8, 14, 34]. The second exploits the *slack* that threads in parallel programs incur when arriving at barriers, either by slowing the first thread to arrive at a barrier [17] or by enabling DVFS once a subset of threads has reached the barrier [25]. The third exploits idle periods during communication between nodes in a cluster to apply DVFS [24] since the processor is not the bottleneck in these scenarios. In addition, some research has considered combinations of these approaches when applying DVFS [9].

Hsu and Feng [14] modeled the effects of DVFS on performance as measured in MIPS (million instructions per second), using regression to derive an application-specific coefficient for the observed performance loss when using a lower DVFS level. This coefficient can then be used to predict the performance when the program runs at other DVFS levels. The authors selected the lowest DVFS level that maintained a given maximum performance loss, thereby maximizing energy savings under a given performance constraint. Similar to their work, we used regression to model the performance impact of applying various levels of DVFS; however, we used multiple hardware events to derive a prediction model rather than a single-performance metric.

Ge et al. [11] compared three strategies for identification of the optimal DVFS level. Specifically, they performed evaluations using (1) CPUSPEED, a system monitor available in Linux Fedora Core that selects the DVFS level based on past CPU (central processing unit) usage; (2) command line control based on a user-estimated DVFS level; and (3) calls to a DVFS library to set the level from within the application. Over a variety of benchmarks and microbenchmarks, the authors found that all approaches were able to save considerable energy, with program-internal control performing the best after extensive optimization.

Isci et al. [15] presented a similar analysis of DVFS. They further found that to meet a given power budget on a CMP (chip multiprocess), DVFS must be applied adaptively based on changes in execution properties to maximize energy savings and performance. The authors also presented an approach to scale DVFS per core independently to meet energy constraints in multiprogram workloads. To optimize throughput and fairness, they utilized performance predictions at different DVFS levels; however, they applied the same static scaling of performance at different DVFS levels regardless of program properties (e.g., memory boundedness), leading to errors in predicted performance. We improved on their work through the use of several application characteristics in our prediction models.

Springer et al. [29] combined DVFS with the deactivation of nodes within a cluster to reduce energy consumption and stay within a specified energy budget. The authors used regression-based prediction of performance under different combinations of DVFS level and number of active nodes, called schedules. Once predictions are made for all schedules, the single schedule that minimizes execution time while not violating the energy constraint is selected for use. Their method requires a relatively large number of sample executions to train the predictor for any given application.

Sasaki et al. [27] implemented a method for performance prediction at various levels of DVFS. Their model is based on multiple linear regression on performance counters collected at run time: They identified the level with the lowest power consumption that met a specified performance requirement and used that for each phase in a program. The model they presented closely followed our prediction model for both concurrency throttling and DVFS, which was presented previously [6]. Furthermore, our prediction model can predict the collective effects of applying both DCT and DVFS (see Section 4.4).

Other related work has focused on deriving an analytical model of DVFS. Ge and Cameron [10] developed a model of speedup achieved by increasing CPU frequency on processors with DVFS. Their model also predicts the effects of increasing the number of cluster nodes used in conjunction with DVFS, although it requires execution samples on all possible numbers of nodes in parallel runs and on all possible processor frequencies in a sequential run. Their analytical model provides considerable insight into the architecture.

4.2.2 Dynamic Concurrency Throttling

Concurrency throttling enables adaptive execution of multithreaded programs in multiprogramming environments. Anderson et al. [1] presented scheduler activations, a technique that lets applications schedule their own threads onto processors and modifies the operating system to allocate processors to applications on demand. In scheduler activations, applications notify the kernel to request more or fewer processors at run time. Similarly, Tucker and Gupta [31] proposed a scheme to give the operating system a mechanism to control the number of processors allocated to each application under multiprogramming, with the application adjusting its concurrency as a result. Hall and Martonosi [12] proposed an approach to identify limited scalability within phases of a parallel application, which they attributed to poor parallelism and false sharing between processors, with the goal of improving the throughput of multiprogram workloads. None of these approaches leaves processors intentionally idle for performance gains or power savings. They rather reallocate processors between applications to improve system throughput.

Stand-alone programs can benefit from concurrency throttling across phases with different scaling and execution time characteristics. For example, it has been observed that simultaneous multithreading (SMT) as implemented on intel hyperthreaded processors tends to perform poorly with parallel scientific applications due to interference between threads in the cache and on other shared execution resources [16, 35]. To address this issue, Zhang et al. [35] developed a loop scheduler that decides the number of threads to use per processor by sampling each possibility in an iterative loop execution scheme.

Adaptive serialization was explored by Voss and Eigenmann [33]; the authors attempted to identify parallel regions that will scale poorly when parallelized and then execute those regions sequentially. They performed identification of poorly scaling regions in two ways. First, if loop length is below some threshold, the loop is serialized because the benefits of parallel execution will be unlikely to outweigh parallelization overhead. Second, they made a prediction of sequential performance based on observed scalability properties of the machine and loop execution times recorded off-line. If the target parallel loop is expected to incur performance loss through parallelization, then it is serialized. Our concurrency throttling approach has the advantage that we do not limit possible concurrency levels to two options. We allow any number of threads between one and the maximum to be used as well.

Jung et al. [16] presented a concurrency throttling method that uses both serialization and SMT disabling. In this work, the authors presented approaches for static and dynamic serialization of parallel regions. Further, they compared two strategies for identification of the optimal number of threads to use per SMT processor. In the first strategy, the two possible configurations—using one or two threads per processor—were simply tested via iterative execution. In the second, they used execution metrics collected while running the program with one thread per processor to predict performance when the program ran with two threads per processor.

Recent work by Li and Martínez [22] considered applying concurrency throttling and DVFS on single CMPs simultaneously, using search algorithms to explore the optimization space. This research had the same objectives as our work; however, the suggested solutions differ significantly. First, our approach is implemented on a real system, rather than a simulated one, verifying that our technique works in practice with all system overheads considered. Second, we utilize performance prediction rather than empirical search methods of the configuration space to reduce the number of test executions necessary to perform adaptation. Finally, our approach targets multiprocessor systems with multicore processors where the combined energy consumption of processors plays a more significant role than in the uniprocessor multicore systems simulated in the other work [22].

Suleman et al. [30] investigated the potential of concurrency throttling to improve both performance and power consumption of parallel applications in shared memory environments [30]. In their work, the authors introduced a technique called *feedback-driven threading* that seeks the optimal number of threads in the face of limited application scalability. The authors considered only bus contention and frequent synchronization as detrimental to parallel scalability, although they acknowledged that other factors—such as cache contention—are likely to be important as well. In our work, we consider any architectural factor that can be captured using a hardware event rates to predict application scalability. Further, we experiment with *real* systems as opposed to the simulated systems used by Suleman et al. [30].

Programmers have long had the ability to specify concurrency levels manually; however, few run-time systems provide the functionality to manage these decisions automatically from within. We present a complete implementation of a run-time system that implements automatic concurrency throttling based on performance predictions of each configuration with the goal of simultaneously improving performance and power consumption in OpenMP applications.

4.3 SCALABILITY PREDICTION

4.3.1 Static Scalability Prediction Models

Performance prediction is a mature research area; therefore, we focus our attention on work most relevant to our performance prediction model for multicore systems. Performance prediction of parallel programs has been studied in great depth, but the majority of research targets off-line prediction. The work most closely related to ours is that of performance prediction for microarchitectural exploration, where the performance of different hardware configurations is predicted through the analysis of quantitative architectural properties, such as reorder buffer size and cache size, as well as application characteristics. The similarities arise from the fact that we are also predicting the effects of architectural changes—specifically the number of processors, cores, and CPU frequency—however, we perform our prediction *online*, during program execution.

Minimizing design space exploration time for processor development has spurred much research on predicting the performance effects of altering various microarchitectural parameters because of the considerable design time reduction that such predictions can provide. Lee and Brooks [20] employed a regression-based prediction strategy with inputs of architectural and application-specific properties. Their goal was to reduce simulation time by limiting the number of required simulations and predicting the performance of unsimulated configurations based on statistical inference of the effects of varying architectural parameters.

In addition to regression analysis, related research on performance prediction uses machine learning through artificial neural networks (ANNs). Singh et al. [28] predicted the performance of two different HPC (high-performance computing) applications with performance variations that are poorly understood on a specific architecture with varying inputs and number of processors. Their prediction model uses multilayer neural networks. They performed training by specifying the input size and the observed execution time to create a function of input size that predicts future queries on how execution time will change with input size. Their approach predicts the performance of the studied applications with an average error as low as 2%; however, all training is application specific; therefore, it can not be reused for new applications or applied for online prediction.

Lee et al. [21] compared the effectiveness of piecewise polynomial regression and ANNs for predicting performance of applications with varying input parameters. Their findings suggest that prediction accuracy was comparable between the two approaches, but each approach was advantageous in different contexts. They reported that the training process was significantly simplified through the use of ANNs. However, they also found that a linear regression model similar to the one that we propose reduced the end-user burden during model specification while still achieving high accuracy.

Carrington et al. [3] demonstrated a framework for predicting the performance of scientific applications in LINPACK and an ocean-modeling application. Their automated approach relies on a convolution method representing a computational mapping of an application signature onto a machine profile. Simple benchmark probes create machine profiles, and a separate tool generates application signatures. They require generating several traces but deliver predictions with error rates between 4.6% and 8.4%, depending on the sampling rates of the underlying traces and platform.

4.3.2 Dynamic Scalability Prediction Models

We present online performance predictors that estimate performance in response to changing DCT and DVFS levels. We refer to each combination of frequency and concurrency configuration available on the system as a hardware configuration, or simply a *configuration*. The predictors use input from execution samples collected at run time on specific configurations to predict the performance on other, untested configurations. We estimate performance for each phase in terms of useful instructions per second, or *uIPC*, which is the IPC computed after omitting instructions used for parallelization or synchronization. By using the *uIPC* prediction, we exploit opportunities to save power by scaling memory-bound phases of computation. Through this scaling, we aim at reducing contention and exploiting slack due to memory or parallelization stalls. The input from the sample configurations consists of the useful IPC ($uIPC_s$) and a set of n hardware event rates $e_{(1..n,s)}$ observed for the particular phase on the sample configuration s. Each event rate $e_{(i,s)}$ is calculated as the number of occurrences of event i divided by the number of cycles elapsed during the execution of the phase with configuration s. The model predicts *uIPC* on a given target configuration t, which we call $uIPC_t$. We refer to this model as *DPAPP* for dynamic phase-aware performance predictor.

Our model for performance prediction differs from previous work in several aspects. The approach works at run time during the execution of an application, and this property limits the computational overhead associated with prediction. To the best of our knowledge, no prior work has considered online performance predictors on shared-memory architectures using run-time input other than execution time. Furthermore, our model adds knowledge to the predictor by accounting for the effects of events in a multiplicative way on the baseline IPC of the sample configuration, rather than an additive way that is often assumed. This means that events are modeled to affect the resulting IPC by some percentage rather than by a fixed amount.

4.3.2.1 Baseline Prediction Model

Our prediction model uses $uIPC_s$ to estimate the effect of the observed event rates that produce the resulting value of $uIPC_t$. The event rates capture the utilization of particular hardware resources that represent scalability bottlenecks, thereby providing insight into the likely impact of hardware utilization and contention on scalability. Although the model can include multiple sample configurations, we begin by describing the simplest case of a single sample and build up the model from there. We model the scalability of $uIPC_t$ as a linear function:

$$uIPC_t = uIPC_s \cdot \alpha_t(e_{(1..n,s)}) + \epsilon_t \tag{4.1}$$

Equation 4.1, reflects the dependence of the function α_t and the constant term ϵ_t on the particular target configuration. We model each target configuration t through coefficients that capture the varying effects of hardware utilization at different degrees of concurrency, mappings of threads to cores, and DVFS levels. In effect, $\alpha()$ scales up or down the observed $uIPC_s$ on the sample configuration based on the observed values of the event rates on the same configuration to estimate the actual $uIPC_t$ on any of the target configurations. The observed event rates determine how much to scale $uIPC_s$, as a linear combination of the sample configuration event rates as

$$\alpha_t(e_{(1..n,s)}) = \sum_{i=1}^{n}(x_{(t,i)} \cdot e_{(i,s)} + y_{(t,i)}) + z_t \tag{4.2}$$

The model's intuition is that changes in event rates indicate varying resource utilization and contention, resulting in either positive or negative effects on $uIPC_s$. The model represents these effects through positive or negative coefficients. While the relationship between event rates and $uIPC$ may not be strictly linear, a simple linear model works well in practice [5, 18, 26]. We estimate the specific coefficients through multivariate linear regression (Section 4.3.2.3). By using an empirical model, we simplify the model retraining required for new architectures since we automatically infer the model from a set of training samples rather than through a detailed architectural description. We combine Equations 4.1 and 4.2 to derive the following equation for $uIPC$ on a particular target configuration t as

$$uIPC_t = uIPC_s \cdot \sum_{i=1}^{n}(x_{(i,t)} \cdot e_{(i,s)}) + uIPC_s \cdot \gamma_t + \epsilon_t \qquad (4.3)$$

Therefore, estimating the value of $uIPC_t$ is equivalent to the proper approximation of the coefficients $x_{(i,t)}$, the constant term ϵ_t, and γ_t. The γ_t variable is the sum of a collection of terms from α_t that represent a coefficient for $uIPC_s$ itself and is independent of the values of $(e_{(1..n,s)})$. This coefficient is defined as $\sum_{i=1}^{n}(y_{(t,i)}) + z_t$.

4.3.2.2 Model Extensions

While the baseline prediction model can be effective for DCT [5], we refine it to improve model accuracy and extend it to predict performance with multidimensional input. Our first extension models $uIPC_t$ as a linear combination of multiple sample configurations from the configuration space. In the context of DVFS and DCT, each sample configuration uses a different number of threads bound to different execution units—cores, processors, or hardware threads—in the machine, at potentially different voltage and frequency levels. Each sample configuration provides some additional insight into execution on other, untested configurations. The use of multiple samples allows the model to "learn" more about each program phase's execution properties that determine performance on alternative configurations. The actual selection of samples can be statistical (e.g., uniform) or empirical (i.e., using some architectural insight such as the number of cores per processor or the number of cores sharing an L2 cache). Equation 4.4 presents the model extended to two samples, with an additional term λ to capture interaction between samples that we describe next.

$$\begin{aligned} uIPC_t &= uIPC_{s1} \cdot \alpha_{(t,s1)}(e_{(1..n,s1)}) + uIPC_{s2} \cdot \alpha_{(t,s2)}(e_{(1..n,s2)}) \\ &\quad + \lambda_t(e_{(1..n,S)}) + \epsilon_t \end{aligned} \qquad (4.4)$$

Using multiple samples allows us to analyze the relationship between each configuration. We include an interaction term for the product of two events in the linear model to capture the relationship statistically. For simplicity, we only consider possible interactions between the same event across multiple configurations, including the product of $uIPC$ on each sample configuration. Our model considers the interplay between multiple configurations. Specifically, we define the interaction term for a model using two samples as

$$\lambda_t(e_{(1..n,S)}) = \sum_{i=1}^{n}(\mu_{(t,i)} \cdot e_{(i,s1)} \cdot e_{(i,s2)})$$
$$+ \mu_{(t,IPC)} \cdot uIPC_{s1} \cdot uIPC_{s2} + \iota_t \qquad (4.5)$$

The interaction term λ_t linearly combines the products of each event across configurations, as well as that of $uIPC$. In Equation 4.5, μ is the target configuration-specific coefficient for each event pair, and ι is the event rate independent term in the model.

On architectures with large and complex configuration spaces, we may need to use additional sample configurations. We can extend our model to an arbitrary collection of samples S of size $|S|$, as

$$uIPC_t = \sum_{i=1}^{|S|}(uIPC_i \cdot \alpha_{(t,i)}(e_{(1..n,i)})) + \lambda_t(e_{(1..n,S)}) + \epsilon_t \qquad (4.6)$$

Using more samples not only generally increases model accuracy but also increases sampling overhead. We address the selection of S in terms of specific configurations as well as its size in Section 4.3.2.6.

We generalize the term λ_t further to account for the interaction between events across $|S|$ samples as

$$\lambda_t(e_{(1..n,S)}) = \sum_{i=1}^{n}(\sum_{j=1}^{|S|-1}(\sum_{k=j+1}^{|S|}(\mu_{(t,i,j,k)} \cdot e_{(i,j)} \cdot e_{(i,k)})))$$
$$+ \sum_{j=1}^{|S|-1}(\sum_{k=j+1}^{|S|}(\mu_{(t,j,k,IPC)} \cdot uIPC_j \cdot uIPC_k)) + \iota_t \qquad (4.7)$$

To further improve model accuracy, we apply variance stabilization in the form of a square-root transformation of the data to reduce correlation between residuals and fitted values, following Lee and Brooks [20]. That is, we take the square root of each term, as well as the response variable, before applying the model. This process results in a more accurate model by reducing model error for the largest and smallest fitted values and causing residuals to more closely follow a normal distribution.

4.3.2.3 Off-line Model Training

We use multivariate linear regression on phases from a set of training benchmarks to approximate coefficients in our model. We record $uIPC$ and a predefined collection of event rates while executing each training benchmark's phases on all configurations. We use multiple linear regression on these values to learn patterns in the effects of sample configuration event rates on the resulting *uIPC* of the target configuration, with each phase's data serving as a training point. Specifically, *uIPC*, the product of IPC and each event rate, and the interaction terms on the sample configurations serve as *independent variables*. *uIPC* on each target configuration serves as the *dependent variable*, in accordance with the equations presented. We develop a model separately for each target configuration, deriving sets of coefficients independently. We select the set of training benchmarks to include variation in properties such as scalability and memory boundedness.

Testing all sample and target configurations off-line for training purposes may become time consuming on architectures with many processing elements and multiple layers of parallelism. To combat this, we prune the target configuration space, using insight on the target system architecture. Specifically, we eliminate symmetric cases in thread binding as well as unbalanced bindings of threads. On emerging architectures that feature hundreds of cores, it may become necessary to further reduce the search space during model training to limit off-line overhead, for example, by uniform sampling of the configuration space used for training. At current multicore system scales, the training process for a fully automated system using our model takes on the order of hours and scales up linearly with the number of configurations.

4.3.2.4   Event Selection Process

The model requires feedback from hardware event rates to predict performance across configurations accurately. Therefore, we must identify specific event rates that result in high prediction accuracy. Unfortunately, the specific events that reflect the performance impact of power knob settings more accurately are not always obvious and depend on the target hardware platform and on workload characteristics. To select the events to use with our model, we use correlation analysis to determine which event rates on the sample configuration are most strongly correlated with the target IPCs. We generate a sorted list of events from higher to lower correlation and select the top *n* events from the sorted list. We determine the number of events to use, *n*, based on how many event registers are available on the target processor architecture. The event selection process is statistical and automated and therefore portable across systems.

4.3.2.5 Predicting Across Multiple Dimensions

We apply our model to predict the performance effects of DCT and DVFS independently or across simultaneous changes in the settings of both power knobs. To predict for simultaneous changes, we collect samples at points along the two-dimensional space by varying the configuration along each prediction dimension. While we could predict along one dimension at a time by selecting the optimal configuration in each dimension sequentially, predicting along both dimensions simultaneously avoids blind spots in the predictions. The first strategy only predicts along the second dimension at the decided optimal level of the first dimension, whereas the second strategy is more likely to find the globally optimal configuration along both dimensions since it considers all combinations on both dimensions. We can generalize the model to predict performance in configuration spaces of higher dimensionality and prune the space through uniform or other sampling schemes that reduce training overhead. Since we predict along two dimensions at most for the purposes of power-performance adaptation, we do not discuss generalization of our model further.

4.3.2.6 Selecting Sample Configurations

Every configuration used as a sample provides additional insight on what the performance will be on a different configuration. Although in principle we can uniformly sample configurations to reduce training and run-time search overhead, some configurations reveal specific architectural bottlenecks to scalability and performance. That is, certain configurations provide further insight into utilization of critical resources such as shared caches and memory bandwidth; therefore, they are stronger predictors than others. We consider architectural properties while selecting the configurations to best serve the prediction model.

When predicting along a single dimension (i.e., concurrency or DVFS level), we use a single sample configuration at the maximum concurrency or frequency available [5]. When predicting along multiple dimensions, our experimental evidence suggests that effective samples are drawn by sampling at points along each dimension. In more detail, we first sample at the maximum concurrency and frequency and then select additional samples, guided by architectural intuition, to improve coverage along each dimension. Each additional sample tests new points along all dimensions. For example, along the concurrency dimension of a four-core system, the first sample uses all four cores at full frequency, and the second sample uses two cores at a different frequency level (thereby providing insight into changes of both the concurrency and frequency dimensions). This technique allows us to limit the number of samples while providing input along each dimension.

4.3.3 Evaluation

In our evaluation, we use a system with two Intel Xeon E5320 quad-core processors. Each core operates at a maximum frequency of 1.86 GHz, and software can reduce this frequency to 1.60 GHz. The system contains 4 GB of memory and runs Linux kernel version 2.6.22. We use seven benchmarks from the NAS suite compiled at class size B (BTC [block tri-diagonal solver], CG [conjugate gradient], FT [discrete 3D Fast Fourier Transform], IS [integer sort), MG [multi-grid on a series of meshes], SP [scalar penta-diagonal solver], and UA [unstructured adaptive mesh]).

Since different mappings of a set of threads to cores may yield significant performance variation, we differentiate between configurations that use the same number of cores but different topologies. On our experimental platform, each of the two processors has two dies, and each die has two cores and an L2 cache bank accessible by both cores on the die. We differentiate between three potential mappings of threads to cores on this system: (1) two threads running in the same die and sharing a common 4 MB L2 cache bank; (2) two threads placed on different dies on the same processor, with private L2 cache banks, using only half of the available memory bandwidth due to their placement on a single processor; and (3) two threads placed on different processors with private L2 cache banks and the full processor memory bandwidth available to them. We specify each hardware configuration using $(X, Y [, Z])$ to denote execution with X processors, Y cores per processor, and DVFS level Z. We use the notation $2s$ to indicate a shared cache and $2p$ to indicate a private cache.

We select a single benchmark to train the model, trading potentially higher prediction accuracy for less training time. Specifically, we use NAS-UA to perform training. UA has over 50 parallel execution phases and widely varying execution characteristics on a per phase basis, including IPC, scalability, locality, and granularity. We select sample configurations for each model to maximize the amount of information available to the model. For the DVFS model, we select a single sample at maximum frequency within any given mapping of threads to cores. We select two samples for DCT: (1,3) and (2,4). Finally, for the unified DVFS-DCT model, we select three samples: $(1,2p,2)$, $(2,2s,1)$, and $(2,4,2)$. These sample configurations are selected as outlined in Section 4.3.2.6 to provide data along each dimension

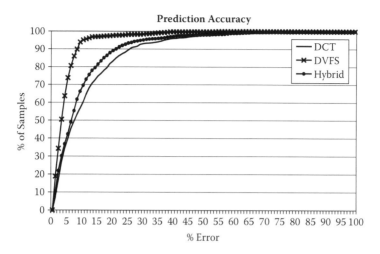

FIGURE 4.1 Cumulative distribution functions of prediction accuracy of the three prediction models.

of adaptation. In all cases, we make predictions for all nonsampled configurations.

Using many events can benefit the model; however, current architectures severely limit the number of events that can be recorded simultaneously, while event multiplexing has significant overhead and questionable accuracy. We set the number of events used in our model to the maximum number of events that the hardware can monitor simultaneously without multiplexing. On our experimental CMP platform, only two event registers are available, and one must always be used to collect *uIPC*, which is mandatory in our model. The statistically selected auxiliary event with the highest correlation with target IPC in the training data is L1 data cache accesses. We derive the model coefficients off-line using linear regression on samples of event rates and *uIPC* on each configuration from the training benchmark.

Figure 4.1 shows the percentage of predicted samples for each model with error less than a particular threshold indicated on the *x*-axis. The results demonstrate high accuracy of the model in all three cases. In particular, the DVFS model yields a median error of 3.0% (4.2% mean), the DCT model a median error of 7.3% (11.2% mean), and the unified model a median error of 6.1% (9.5% mean). We note that prediction is performed with input from one, two, or three sample configurations for all remaining configurations. The higher accuracy of predicting DVFS than DCT indicates that DVFS has simple, mostly linear, effects on performance that our model captures. DCT, on the other hand, has more complex and often nonlinear performance effects due to irregular, nonmonotonic scalability patterns in many parallel execution phases. Of the 20 possible configurations, the unified model correctly identifies the single best configuration in 35% of cases. The model incorrectly selects one of the ten worst configurations in only 7% of cases.

4.4 MULTIDIMENSIONAL POWER-PERFORMANCE ADAPTATION

We consider the synergistic integration of DVFS and DCT. We describe a methodology for dynamically adapting the execution of parallel applications on shared-memory multicore systems to reduce power consumption while sustaining or improving performance. We also present a run-time system called ACTOR that adapts program execution to enforce concurrency and voltage-/frequency-level decisions without a priori knowledge of application characteristics. ACTOR uses the DPAPP model for scalability prediction presented in the previous section. We evaluate the complete framework (ACTOR, DPAPP) in the context of simultaneous DVFS and DCT optimization on a real multiprocessor with multicore processors.

4.4.1 Scalability Analysis of Parallel Applications

To motivate DCT, we briefly analyze the scalability of a parallel benchmark suite on our experimental platform (see Section 4.3.3 for the specifications of this platform). To conduct this evaluation, we execute each benchmark under all nonsymmetric configurations on the experimental hardware and record execution time and energy consumption. Figure 4.2 presents the results. Our hardware testbed has ten possible nonsymmetric configurations. In all experiments, we measure full system energy per run using a Watts Up Pro power meter. We also compute average power consumption from execution time and total energy consumption.

The results show that, in principle, parallel benchmarks do not scale perfectly on the target hardware. In particular, only one benchmark achieves best performance when using all eight cores. We observe three patterns of scalability in our experiments. First are those benchmarks that manage reasonable speedup through the utilization of additional cores (BT, FT, and UA). Second are benchmarks that incur nonnegligible performance *loss* when using *more* cores (IS and MG). Third are benchmarks that neither gain nor lose performance from higher concurrency (CG and SP). Despite poor utilization of additional cores, energy consumption generally increases

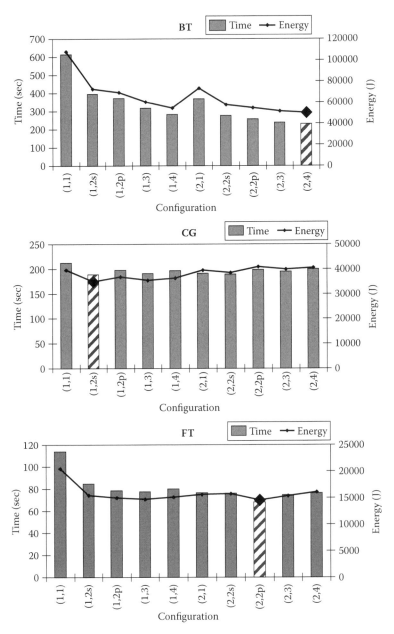

FIGURE 4.2 Execution time (bars) and energy consumption (lines) of the benchmarks across all configurations. The configurations with the best performance and energy for each benchmark are marked with stripes and a large diamond, respectively. *(continued)*

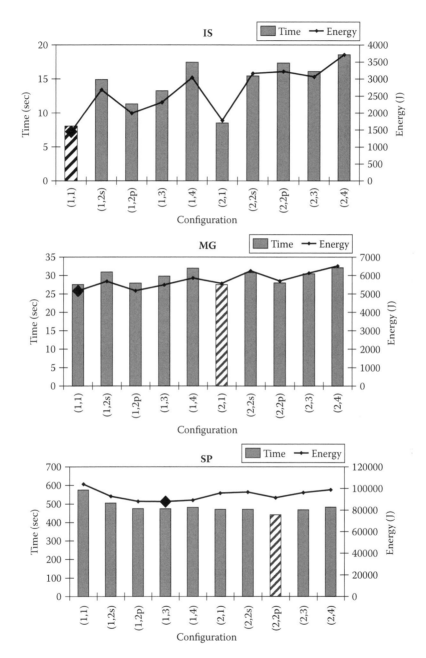

FIGURE 4.2 *(continued)*. Execution time (bars) and energy consumption (lines) of the benchmarks across all configurations. The configurations with the best performance and energy for each benchmark are marked with stripes and a large diamond, respectively. *(continued)*

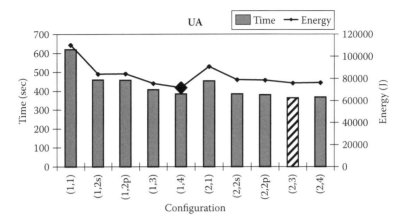

FIGURE 4.2 *(continued)*. Execution time (bars) and energy consumption (lines) of the benchmarks across all configurations. The configurations with the best performance and energy for each benchmark are marked with stripes and a large diamond, respectively.

with more cores. It is predominantly the second category of benchmarks (negative scalability) that motivates the use of DCT to throttle concurrency to more efficient levels, although in many cases the third category (flat scalability) can also be optimized by reducing power consumption while sustaining performance.

The most energy-efficient configuration coincides with the most performance-efficient configuration for four of the seven benchmarks (BT, CG, FT, and IS). For three benchmarks (MG, SP, and UA), the user can use fewer than the performance-optimal number of cores to achieve substantial energy savings, with a marginal performance loss. We also observe that for a given number of threads, performance can be very sensitive to the mapping of threads to cores (e.g., BT, FT, and SP when executed with two or four threads). Even if performance is insensitive to the mapping of threads to cores, power can be sensitive to the mapping of threads to cores. In MG, for example, placing two threads on two dies on the same processor is imperceptibly less performance efficient but significantly more energy efficient than placing the two threads on different processors.

We attribute poor scalability of several benchmarks to memory contention at all levels of the memory hierarchy. Two cores sharing a cache rarely benefit from this sharing. They suffer instead from destructive interference, which manifests as more conflict misses between threads. It is also interesting to note the substantial performance benefit seen through execution with two threads per processor and private cache space for each thread. Additional threads increase the demand for memory bandwidth and produce contention on the shared front-side bus. These issues combine to limit scalability most in applications that are memory bound and have primary or secondary working sets that are too large to fit in the cache. The results demonstrate opportunities to use fewer threads to improve performance and energy consumption.

4.4.2 Run-Time Support for Energy-Efficient Multithreading

In this section, we present a phase-aware concurrency throttling algorithm on our hardware testbed. We further demonstrate an adaptive run-time system that responds to observed program behavior to adapt execution, called *ACTOR* (for adaptive concurrency throttling optimization run-time system). We then evaluate the potential of the algorithm for reducing power and energy. Section 4.4.3 describes the way in which ACTOR is extended to simultaneously adapt DCT and DVFS.

Scientific codes are dominated by iterative execution of phases. ACTOR exploits this property to sample hardware event rates in the first few phase traversals and set the concurrency of each phase to the predicted optimal operating point for the rest of phase traversals. ACTOR strives to sample and adapt early during execution of the program. Live search of the optimization space for operating points of concurrency can also be performed by timing phases at different configurations and running search heuristics such as greedy hill-climbing [5, 22] or simulated annealing [19]. However, as the number of feasible hardware configurations increases with the introduction of more cores per processor, direct search methods may spend significant portions of the execution time sampling suboptimal configurations rather than actually optimizing the program. This disadvantage becomes apparent in codes that traverse dominant parallel phases only a few times.

Even if direct search methods are used for off-line autotuning by repetitive executions of the entire program [2], searching the optimization space for any input on any feasible configuration of processing units may be prohibitive. ACTOR prunes the search space for concurrency optimization to a constant number of samples. The DPAPP-based concurrency throttling algorithm takes as a parameter the list of sample configurations. The number of samples S corresponds to the number of times each phase needs to be executed before deriving a prediction for the optimal operating point and is used to control sampling overhead. In our prototype implementation of DCT, we use a sample rate of $S = 2$.

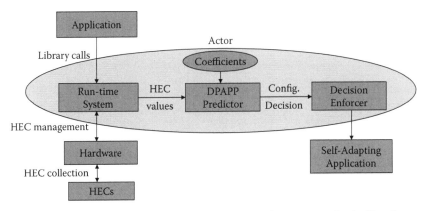

FIGURE 4.3 The structure of the ACTOR run-time system including interface to hardware event counters (HECs).

The structure of the ACTOR system is shown in Figure 4.3. The controller is dynamic, in the sense that it adapts the program as it executes, with no prior knowledge of program characteristics. ACTOR requires simple, formulaic code instrumentation and estimates optimal operating points of concurrency using samples of critical hardware event rates from live executions of program phases. Specifically, the library controls the first S phase traversals to execute on the desired sample configurations and collect event rates. At the end of the sampling period, collected event rates are used by DPAPP to predict the *uIPC* of each phase on nonsampled configurations. Once predictions for a phase are obtained, all subsequent traversals of this phase are executed at the predicted optimal configuration. ACTOR enforces configuration decisions through the Linux processor affinity system call, $sched\_setaffinity()$, and multithreading library-specific calls for changing concurrency levels, such as $omp\_set\_num\_threads()$ in OpenMP. The overhead of using ACTOR in terms of the time spent executing library code is approximately 500,000 cycles per program phase (250 μs on a 2-GHz processor), which is negligible for realistic applications.

Certain assumptions are necessary to implement concurrency throttling in our system, and we outline those in the following discussion. First, we rely on the capability of the run-time system to change the number of threads used to execute a phase of parallel code at run time. This capability is available in OpenMP, at the granularity of parallel loops and parallel regions. However, changing the number of threads at run time may not be possible in some applications due to (1) data initialization that depends on the number of threads used or (2) alternative control paths during parallel execution, the choice of which depends on the thread ID. Second, the phases of an application must be executed at least S times to complete the sampling stage. Finally, the workload of each phase between executions must remain relatively stable. In practice, this is the case in most regular and irregular scientific codes.

While we have specifically designed ACTOR for use with iterative scientific applications, our approach applies to other categories of applications as well. The basic principle of ACTOR can be applied with any definition of a program phase where concurrency can be dynamically adjusted. For example, in noniterative, synchronization-intensive, or heterogeneous multithreaded codes, if an existing phase identification technique can be employed to identify repetitive behavior where concurrency is modifiable, ACTOR can be applied for DCT. For server workloads, the application may be treated as one phase, and a limited timeframe can be monitored to decide concurrency for the application.

Energy savings using DCT come through two paths: first, by reducing execution time since energy consumption is reduced proportionally in this case; and second, through the deactivation of processing units, which reduces power consumption. The power consumption of a processing unit depends on utilization of its functional units since clock gating limits the power dissipation of functional units when they are idle. Furthermore, a processor can be transitioned to a low-power mode when it is not in use. For example, on Intel Pentium 4 processors, the *hlt* instruction transitions the processor to a low-power mode, where power consumption is reduced from approximately 9 W when idle to 2 W when halted. Recent Intel processors possess multiple C states in which the processor can enter increasingly deep levels of sleep to conserve additional power, at the cost of increased latency to return to operational mode. While we do not manually control the transitioning between power states of the processors from within the run-time system, the operating system does so when the processor remains idle for a configurable time period. We have experimentally verified that in Linux 2.6 kernels, the operating system transitions processors to the halted state and keeps them in this state during approximately 90% of the time they are idle.

4.4.3 Integrating DCT with DVFS

We extend the ACTOR framework to dynamically adjust multiple power-performance knobs [7]. The advantage of a hybrid adaptation approach is that it enables the system to capitalize on the benefits of each knob's benefits.

Multidimensional power-performance adaptation within ACTOR operates in the same way as DCT adaptation alone. Just as with DCT adaptation, ACTOR controls the first few traversals of each program phase to run on predetermined configurations, characterized by the number of threads, the placement of threads on cores, and the DVFS level. ACTOR monitors execution properties to serve as input to the DPAPP prediction model. Once predictions are made, ACTOR controls the execution of each phase to use the configuration selected by the DPAPP predictor. Program control of multidimensional adaptation requires no additional instrumentation beyond that already present for DCT. The run-time library is extended also to control DVFS at a per phase granularity.

To perform multidimensional adaptation, we assume that we can set each processor of the system to execute at a voltage/frequency level chosen from a predetermined set through a privileged instruction. We assume global voltage and frequency scaling for each processor as a whole, as opposed to per core, since this is the only option available on our experimental hardware and to better support parallel codes by avoiding load imbalance when setting the frequency of individual cores independently during parallel execution phases. Our modeling methodology does not preclude and can be generalized to per core DVFS schemes. We conduct physical experimentation, using hardware timers and power meters to measure performance and energy, respectively.

When adapting DCT, the library can compare configurations simply using the predicted *uIPC*. However, when considering DVFS alone or in hybrid approaches, the model needs adjustment to ensure valid performance comparisons. A problem arises because at lower frequencies each cycle lasts longer, which causes higher IPCs to occur at lower frequencies while the program actually runs slower. For this reason, the predictor calculates IPC before making comparisons and takes into consideration the frequency levels.

4.4.4   Evaluation

In this section, we evaluate the use of the presented prediction models in conjunction with run-time adaptation of multithreaded scientific codes. We begin by evaluating the use of only DVFS or DCT. We then analyze two schemes for adapting both DVFS and DCT, first applying them sequentially and then in a unified manner. Finally, we compare prediction-based adaptation against empirical search in identifying optimal configurations, using both an exhaustive search and a binary search. Figure 4.4 presents the results of adaptation through these various mechanisms for each benchmark, and Figure 4.5 shows geometric means of normalized energy and execution time.

We compare various adaptation strategies against "static executions" that use a single configuration through the entire run. We specifically compare adaptation strategies to static executions using full concurrency and frequency (*static*) and to the best performing of all static executions (*static optimal*). We derive the static optimal ex post facto: We cannot know the static optimal online without exhaustive off-line execution of each application on all configurations and for each input. We use static optimal as a practically unrealistic baseline for comparison with other strategies. The static optimal is not necessarily the overall optimal, as each phase may have its own optimal configuration. We do not consider this possibility as its identification requires exponential time, making it unrealistic even for off-line use.

4.4.4.1 Single-Dimension Adaptation

In this analysis, we make adaptation decisions by selecting the configuration with the highest predicted performance. The results of applying DVFS support the intuition that DVFS is generally unable to improve performance. The literature includes corner cases of memory-bound phases where this assumption is violated, and scaling down frequency can marginally improve execution time [11], but these phases are rare exceptions. Our experiments reveal no benefit in terms of performance or energy from adapting to the DVFS level with the highest predicted performance. Without tolerating some loss in performance, DVFS alone is not generally able to significantly benefit energy consumption. We can attribute this result to some extent to our system having only two voltage/frequency levels, with the lower level not substantially more power efficient than the higher level (1.6 vs. 1.86 GHz).

Using DCT with the prediction model provides substantial benefits in execution time (9.5% mean savings), power consumption (3.7% mean savings), and energy consumption (13.1% mean savings) compared to static execution with all cores active. Despite the positive result, mispredictions for two benchmarks result in an observed increase in execution time: FT by 11.6% and SP by 4.2%. However, SP still manages energy savings of 2.5% because of reduced power consumed by the fewer active cores, while FT increases energy by only 1.0% and is the only benchmark not to have energy consumption reduced through DCT. In contrast, the largest benefit

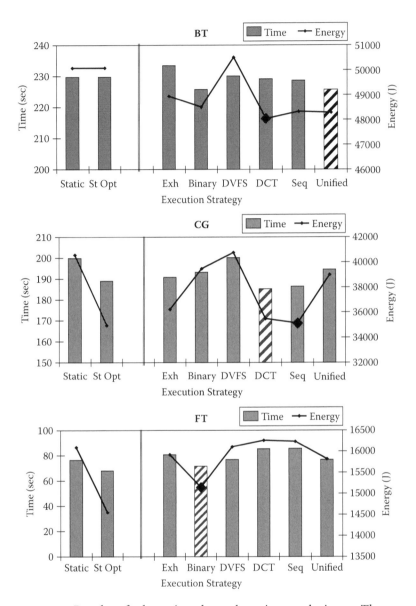

FIGURE 4.4 Results of adaptation through various techniques. The group of bars left of the divider represent static configurations, and those right of the divider are the adaptive strategies. The adaptive configurations with the best performance and energy for each benchmark are marked with stripes and a large diamond, respectively. *(continued)*

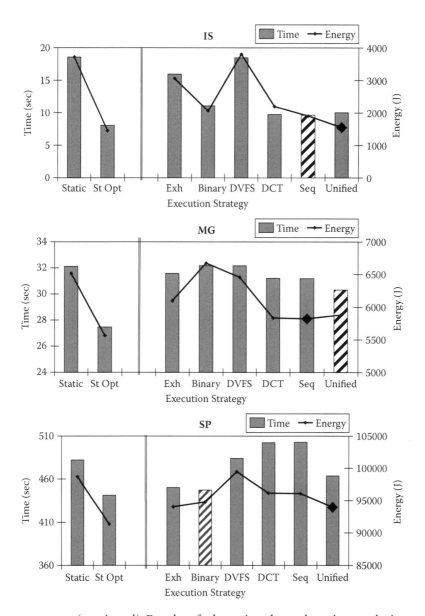

FIGURE 4.4 *(continued)*. Results of adaptation through various techniques. The group of bars left of the divider represent static configurations, and those right of the divider are the adaptive strategies. The adaptive configurations with the best performance and energy for each benchmark are marked with stripes and a large diamond, respectively. and that with the lowest energy consumption with a star.

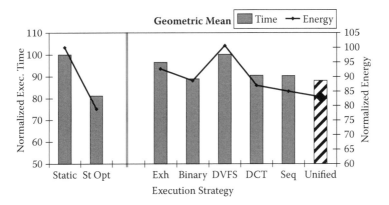

FIGURE 4.5 Geometric means of the benefits of adaptation through various strategies. The adaptive configurations with the best performance and energy for each benchmark are marked with stripes and a large diamond, respectively.

occurs with IS, which sees a 40.7% reduction in energy consumption. When compared to the static optimal execution, DCT is within 8.1% of mean performance and even surpasses that performance with CG (by 2.0%) due to phase awareness.

4.4.4.2 Multidimensional Adaptation

We combine the two power-performance knobs in multiple ways. First, we consider applying the two knobs sequentially. We first apply DCT and then DVFS on the active cores in each phase, as DCT has a clear advantage over DVFS in reducing power while improving performance on our experimental platform. Since we make decisions to maximize predicted IPC and our platform has only two DVFS levels with a small frequency difference, DVFS adds very little benefit to DCT alone, and no reduction in execution time compared to DCT alone occurs. However, DVFS reduces power and energy consumption by 2.3% and 2.0%, respectively, beyond DCT on average, by identifying several phases to reduce frequency without a negative impact on performance.

Second, we apply the unified prediction approach. The major advantage of the unified prediction approach is that it eliminates blind spots in the configuration space during the prediction process. Whereas sequential application of DVFS and DCT will only evaluate DVFS options on the decided DCT level, the unified approach considers all possible values of each parameter in a single step. Furthermore, the unified scheme uses the same number of execution samples as the sequential approach; however, it uses all samples for both DVFS and DCT instead of dividing them between the two. It is able to exploit its higher accuracy to identify more effective DCT levels and finds opportunities to apply DVFS that do not harm performance.

Because of its advantages over the sequential approach, the unified scheme improves performance by 2.2% and reduces energy by 1.9% (geometric mean improvements over the sequential approach). When compared to the default execution using maximum concurrency and frequency, the advantages of unified adaptation become even clearer. Specifically, we see an 11.8% speedup and a 5.9% reduction in power consumption, resulting in an overall reduction in energy consumption of 17.0% (geometric mean improvements over static execution). In fact, all benchmarks exhibit better performance with the exception of FT, which slows by a mere 0.3%. Similarly, all benchmarks benefit from reduced energy consumption. Even when compared to the oracle-derived executions on the static optimal configuration for each benchmark, unified adaptation achieves energy consumption within 4.2% and performance within 7.0% (geometric means). In one case, BT, unified adaptation achieves better performance (by 2.0%) than the static optimal due to identification of improved configurations in several isolated phases. This indicates that prediction models are both viable and effective in addressing the multidimensional program adaptation problem.

4.4.4.3   Prediction versus Search Approaches

We also evaluate two empirical search approaches to identify optimal DVFS and DCT configurations. The first of these performs an exhaustive search of the configuration space before making a decision while measuring execution time of phases with each configuration. This approach does not require off-line training; therefore, the programmer can use it with minimal effort. However, the online overhead of testing many possible configurations stands to reduce the potential benefit of adaptation considerably, which is what occurs in practice. The exhaustive search method reduces execution time by 3.5%, power by 4.0%, and energy by 7.3%, well below the savings of prediction-based techniques. Exhaustive search is superior to prediction schemes in SP only. SP executes 400 workload-invariant iterations.

For purposes of comparison, we also consider a heuristic search approach, based on a binary search of the configuration space, similar to the approach evaluated by Li and Martínez [22]. Our implementation of a binary search begins by executing at full concurrency and frequency. It proceeds with sequential binary searches of the concurrency and DVFS dimensions. During the searches, if a sample is tested with worse performance than the first sample, concurrency or DVFS is increased in the next-tested sample. This approach has considerably reduced overhead compared to exhaustive search because many configurations need not be tested, resulting in 7.6% better performance and 4.1% lower energy consumption (geometric mean improvements over exhaustive search). Compared to static execution, performance is improved by 11.1% and energy by 11.4%. This suggests that a heuristic search can be effective in the context of adapting DCT and DVFS at run time. However, it still falls short of the static optimal configuration by 7.7% for performance and 9.8% for energy.

Comparing the unified prediction model to a binary search provides further insight into the merits of each approach. A binary search achieves performance that is 0.7% worse than the unified prediction approach while consuming 5.6% more energy (geometric mean differences). A binary search suffers from blind spots that prevent identification of effective configurations at low concurrency or DVFS levels. At these levels, the processor consumes less power. The unified prediction model identifies such levels and can therefore reduce power further (by 5.5% on average). A binary search does achieve better performance than the unified model in three of six cases (CG, FT, and SP); however, energy consumption is higher in all but one case (SP). In particular, a binary search performs poorly for MG and IS, which contain too few iterations to amortize the search overhead, in contrast to BT and SP, for which a binary search excels since these benchmarks execute 200 and 400 iterations, respectively. As future systems increase in parallelism and number of DVFS levels available, we expect the relative benefit of prediction-based adaptation schemes to grow.

4.5 CONCLUSIONS

The number of cores in a single processor is increasing at a nearly exponential rate. By putting additional simple cores in a single chip, processor designers hope to stabilize power consumption while improving performance. Unfortunately, even in the highly specialized HPC domain, software cannot easily achieve strong scaling to many cores on a single chip. Our experiments demonstrated that HPC applications observe performance losses beyond even modest concurrency levels on an eight-core system. Given this observation, we strive for economizing on the number of cores activated on chip to reduce power consumption while sustaining, or even improving, application performance.

In this chapter, we presented a model to predict the performance effects of applying multiple energy-saving techniques simultaneously. The model applies statistical analysis of hardware event rates to estimate how voltage/frequency scaling or DCT influence performance in workload execution phases and across system configurations. Over a range of benchmarks, our model achieved a low median error of 6.1% in prediction in response to simultaneous tuning of DVFS and DCT. The high prediction accuracy allows for the successful identification of efficient operating points and phase-aware adaptation in HPC applications.

We applied our model to adapt program execution by regulating concurrency as well as DVFS levels. Our results indicated that while DVFS on its own is not ideal for the HPC domain where performance is critical, DCT is a promising alternative. Specifically, DCT alone achieved 9.5% performance improvement and 3.7% power reduction on a CMP-based system. Further, we found that combining the two approaches in a synergistic fashion could simultaneously improve performance and energy efficiency relative to either approach in isolation. Specifically, a unified adaptation model achieved performance improvements of 11.8%, power savings of 5.9%, and energy savings of 17.0% compared to using all cores at full frequency. The unified model outperformed techniques that apply DCT and DVFS sequentially instead of simultaneously. We also compared our prediction model to methods using exhaustive or binary search of system configurations by measuring execution time while testing configurations. We found that while a binary search outperformed an exhaustive search, it was not superior to the prediction-based approach due to overhead and blind spots. As we scale to more cores and DVFS levels, the overhead of search-based approaches is likely to increase, thus widening the advantage of prediction. Given that the performance of prediction-based methods can effectively approximate the performance of an oracle, we conclude that they are a viable alternative for future-generation systems with many cores and fine-grain power control capabilities.

REFERENCES

1. T. Anderson, B. Bershad, E. Lazowska, and H. Levy. Scheduler Activations: Effective Kernel Support for the User-Level Management of Parallelism. *ACM Transactions on Computer Systems*, 10(1):53–79, February 1992.
2. K. Asanovic, R. Bodik, B. C. Catanzaro, J. J. Gebis, P. Husbands, K. Keutzer, D. A. Patterson, W. L. Plishker, J. Shalf, S. W. Williams, and Katherine

A. Yelick. *The Landscape of Parallel Computing Research: A View from Berkeley*. Technical report ucb/eecs-2006-183, EECS Department, University of California at Berkeley, December 2006.

3. L. Carrington, A. Snavely, X. Gao, and N. Wolter. A Performance Prediction Framework for Scientific Applications. In *Workshop on Performance Modeling—ICCS*, Melbourne, Australia, June 2003.

4. K. Chakraborty, P. Wells, and G. Sohi. *A Case for an Over-provisioned Multicore System: Energy Efficient Processing of Multithreaded Programs*. Technical Report TR-1607, Department of Computer Sciences, University of Wisconsin–Madison, 2007.

5. M. Curtis-Maury, F. Blagojevic, C. D. Antonopoulos, and D. S. Nikolopoulos. Prediction-Based Power-Performance Adaptation of Multithreaded Scientific Codes. *IEEE Transactions on Parallel and Distributed Systems*, 19(10): 1396–1410, October 2008.

6. M. Curtis-Maury, J. Dzierwa, C. Antonopoulos, and D. Nikolopoulos. Online Power-Performance Adaptation of Multithreaded Programs using Hardware Event-Based Prediction. In *Proceedings of the 20th ACM International Conference on Supercomputing*, Queensland, Australia, June 2006.

7. M. Curtis-Maury, A. Shah, F. Blagojevic, D. Nikolopoulos, B. R. de Supinski, and M. Schulz. Prediction Models for Multi-Dimensional Power-Performance Adaptation on Many Cores. In *Proceedings of the International Conference on Parallel Architectures and Compilation Techniques*, Toronto, Canada, October 2008.

8. V. Freeh, D. Lowenthal, F. Pan, and N. Kappiah. Using Multiple Energy Gears in MPI Programs on a Power-Scalable Cluster. In *Proceedings of the 2005 ACM SIGPLAN Symposium on Principles and Practices of Parallel Programming (PPoPP'05)*, Chicago, IL, June 2005.

9. V. Freeh, F. Pan, D. Lowenthal, N. Kappiah, R. Springer, B. Rountree, and M. Femal. Analyzing the Energy-Time Tradeoff in High-Performance Computing Applications. *Transactions on Parallel and Distributed Systems*, 5(11), 835–848, June 2007.

10. R. Ge and K. W. Cameron. Power-Aware Speedup. In *Proceedings of the 21st IEEE International Parallel and Distributed Processing Symposium*, Long Beach, CA, March 2007.

11. R. Ge, X. Feng, and K. Cameron. Performance-Constrained Distributed DVFS Scheduling for Scientific Applications on Power-Aware Clusters. In *Proceedings of the 17th IEEE/ACM High-Performance Computing, Networking, and Storage Conference (SC'05)*, Seattle, WA, November 2005.

12. M. Hall and M. Martonosi. *Adaptive Parallelism in Compiler-Parallelized Code*. In Proceedings of the 2nd DUIF Computer Workshop, Stanford, CA, August 1997.

13. J. Howard et al. A 48-Core IA-32 Message-Passing Processor with DVFS in 45nm CMOS. In *Proceedings of the International Solid State Circuits Conference*, pp. 108–109, San Francisco, CA, 2010.

14. C. Hsu and W. Feng. A Power-Aware Run-Time System for High-Performance Computing. In *Proceedings of the the ACM/IEEE International Conference on High-Performance Computing, Networking, and Storage (Supercomputing)*, Seattle, WA, September 2005.

15. C. Isci, A. Buyuktosunoglu, C. Cher, P. Bose, and M. Martonosi. An Analysis of Multi-Core Global Power Management Policies: Maximizing Performance for a Given Power Budget. In *Proceedings of the 39th International Symposium on Microarchitecture*, Orlando, FL, December 2006.

16. C. Jung, D. Lim, J. Lee, and S. Han. Adaptive Execution Techniques for SMT Multiprocessor Architectures. In *Proceedings of the Tenth ACM SIGPLAN Symposium on Principles and Practice of Parallel Programming*, Chicago, IL, June 2005.

17. N. Kappiah, V. Freeh, and D. Lowenthal. Just in Time Dynamic Voltage Scaling: Exploiting Inter-Node Slack to Save Energy in MPI Programs. In *Proceedings of IEEE/ACM Supercomputing'2005: High Performance Computing, Networking Storage, and Analysis Conference*, Seattle, WA, November 2005.

18. T. S. Karkhanis and J. E. Smith. A First-Order Superscalar Processor Model. In *Proceedings of the 31st International Symposium on Computer Architecture*, Munich, Germany, June 2004.

19. S. Kirkpatrick, C. Gelatt, and M. Vecchi. Optimization by Simulated Annealing. *Science*, 220(4598):671–680, 1983.

20. B. Lee and D. Brooks. Accurate and Efficient Regression Modelling for Microarchitectural Performance and Power Prediction. In *Proceedings of the 12th International Conference on Architectural Support for Programming Languages and Operating Systems*, San Josp, CA, June 2006.

21. B. Lee, D. Brooks, B. R. de Supinski, M. Schulz, K. Singh, and S. A. McKee. Methods of Inference and Learning for Performance Modeling of Parallel Applications. In *Proceedings of the International Symposium on Principles and Practices of Parallel Programming*, San Jose, CA, March 2007.

22. J. Li and J. Martínez. Dynamic Power-Performance Adaptation of Parallel Computation on Chip Multiprocessors. In *Proceedings of the 12th International Symposium on High-Performance Computer Architecture*, Austin, TX, February 2006.

23. Y. Li, B. C. Lee, D. Brooks, Z. Hu, and K. Skadron. CMP Design Space Exploration Subject to Physical Constraints. In *Proceedings of the IEEE International Symposium on High Performance Computer Architecture*, Austin, TX, February 2006.

24. M. Lim, V. Freeh, and D. Lowenthal. Transparent Frequency and Voltage Scaling of Communication Phases in MPI Programs. In *Proceedings of IEEE/ACM Supercomputing*, Tampa, FL, November 2006.

25. C. Liu, A. Sivasubramaniam, M. Kandemir, and M. Irwin. Exploiting Barriers to Optimize Power Consumption on CMPs. In *Proceedings of the 19th International Parallel and Distributed Processing Symposium*, Denver, CO, April 2005.

26. T. Moseley, J. Kim, D. Connors, and D. Grunwald. Methods for Modelling Resource Contention on Simultaneous Multithreaded Processors. In *Proceedings of the 2005 International Conference on Computer Design*, pp. 373–380, San Jose, CA, October 2005.

27. H. Sasaki, Y. Ikeda, M. Kondo, and H. Nakamura. An Intra-Task DVFS Technique Based on Statistical Analysis of Hardware Events. In *Proceedings of the International Conference on Computing Frontiers*, Ischia, Italy, May 2007.

28. K. Singh, E. Ipek, S. A. McKee, B. R. de Supinski, M. Schulz, and R. Caruana. Predicting Parallel Application Performance via Machine Learning Approaches. *Concurrency and Computation: Practice and Experience*, 19(17): 2219–2235, May 2007.

29. R. Springer, D. Lowenthal, B. Rountree, and V. Freeh. Minimizing Execution Time in MPI Programs on an Energy-Constrained, Power-Scalable Cluster. In *Proceedings of the 11th ACM SIGPLAN Symposium on Principles and Practice of Parallel Programming*, New York, March 2006.

30. M. A. Suleman, M. K. Qureshi, and Y. N. Patt. Feedback-Driven Threading: Power-Efficient and High-Performance Execution of Multi-threaded Workloads on CMPs. In *Proceedings of the International Symposium on Architectural Support for Programming Languages and Operating Systems*, Seattle, WA, March 2008.

31. A. Tucker and A. Gupta. Process Control and Scheduling Issues for Multi-programmed Shared-Memory Multiprocessors. In *Proceedings of the 12th ACM Symposium on Operating Systems Principles (SOSP'89)*, pp. 159–166, Litchfield Park, AZ, December 1989.

32. S. Vangal, J. Howard, G. Ruhl, S. Dighe, H. Wilson, J. Tschanz, D. Finan, P. Iyer, A. Singh, T. Jacob, S. Jain S. Venkataraman, Y. Hoskote, and N. Borkar. An 80-Tile 1.28TFLOPS Network-on-Chip in 65nm CMOS. In *Proceedings of the International Solid State Circuits Conference*, pp. 5–7, San Francisco, CA, 2007.

33. M. Voss and R. Eigenmann. Reducing Parallel Overheads through Dynamic Serialization. In *Proceedings of the 13th International Parallel Processing Symposium and Symposium on Parallel and Distributed Processing (IPPS/SPDP)*, pp. 88–92, San Juan, Puerto Rico, April 1999.

34. A. Weissel and F. Bellosa. Process Cruise Control: Event-Driven Clock Scaling for Dynamic Power Management. In *Proceedings of the 2002 International Conference on Compilers, Architecture and Synthesis for Embedded Systems*, pp. 238–246, Grenoble, France, October 2002.

35. Y. Zhang, M. Burcea, V. Cheng, R. Ho, V. Cheng, and M. Voss. An Adaptive OpenMP Loop Scheduler for Hyperthreaded SMPs. In *Proceedings of PDCS-2004: International Conference on Parallel and Distributed Computing Systems*, San Francisco, September 2004.

CHAPTER 5

Exploring Trade-Offs between Energy Savings and Reliability in Storage Systems

Ali R. Butt, Puranjoy Bhattacharjee, Guanying Wang, and Chris Gniady

CONTENTS

5.1 INTRODUCTION

Digital data are increasingly playing a central role in our lives. Be it family photographs and music, scientific observations and experimentation results, or corporate and government data, digital storage has emerged as the reliable and cost-efficient means for providing high data availability and fast access. However, with improvements in storage media densities that increase disk capacities, the chances of data loss are increasing. Furthermore, employment of a large number of devices to store massive amounts of data increases the probability that some disk failures will occur. Recent works [1–5] have identified many reasons for data loss, such as crash failures that result in disks that become completely inaccessible and latent sector errors (LSEs) by which portions of stored data are corrupted but the disks remain accessible. As users increasingly store valuable, irreplaceable data on computers, the reliability requirements are becoming more stringent. This is evident by the user reaction and uproar to even small data-loss incidents at industry leaders such as at Google [6–8] or Facebook [9]. The emerging cloud computing paradigm [10–14], which decouples users from the system-level details of storage management, is often wrongly interpreted by the users as a failure-proof storage system. Consequently, cloud designers are faced with the extreme challenge of providing near-flawless reliability for the stored data [6–9], and they naturally turn to using a high degree of redundancy in the system to guard against the unexpected.

Employing a large number of storage devices, as is the current trend in realizing large-capacity storage systems, not only leads to an increase in the number of failures in the system [1, 2, 4, 15] but also results in significant energy consumption [15–18] (accounting for as much as 27% of a data center's total energy consumption [17]).

Recent research has shown that the use of energy conservation techniques has both financial and environmental impacts [19–21]. Consequently, many organizations employ active energy management. In fact, it was recently observed at the NSF (National Science Foundation) Science of Power Management Workshop [22] that researchers and IT (information technology) practitioners now expect the periodic operating budgets, especially cost of energy for running large disk farms, to exceed even the initial procurement expenses [23]. Consequently, organizations are faced with the two challenges of providing high reliability and reducing energy consumption of the storage systems. However, addressing these challenges can be conflicting and entail careful consideration for optimum system reliability, performance, and energy efficiency [24].

Given the focus and need for improving both energy efficiency and data reliability, this chapter explores the interactions between the reliability and energy management tasks and novel storage system organization that maximizes energy efficiency and reliability.

The rest of the chapter is organized as follows. In Section 5.2, we look into the problem of reliability and energy efficiency trade-off for disks. In Section 5.3, we discuss the state of the art in techniques to improve reliability and energy efficiency in disks and provide a thorough survey of these techniques. Next, we discuss the concept of the energy-reliability product (*ERP*) in Section 5.4, followed by Section 5.5, where we examine the application of *ERP*. In Section 5.6, we discuss ways to extend the concept of *ERP* to other domains. Finally, Section 5.7 concludes the chapter.

5.2   RELIABILITY VERSUS ENERGY EFFICIENCY OF STORAGE SYSTEMS

Reconciling the apparently conflicting goals of improving reliability and energy efficiency of storage systems entails innovation. We now discuss the motivation for providing tools and technologies that can achieve this goal and the resulting challenges.

Increasing disk reliability reduces opportunities for saving energy and vice versa. A large body of research has explored disk reliability and its performance implications [1–5, 25–29]. One of the commonly used mechanisms for improving data reliability is RAID (redundant array of inexpensive disks) [25], which provides protection against total disk failures. RAID was designed to provide high performance and reliability; subsequently, it has high energy consumption [15–18]. The higher energy consumption arises from data and parity bits being striped across the disks in the system, which results in all disks being active when performing I/O (input/output) operations. While RAID provides protection against device failure, it does not prevent all errors. Most recent disk failures are due to portions (sectors) of a disk going bad [2, 4]. Such sectors may not be read and discovered to be corrupted in time when they can be fixed, leading to eventual data loss.

To fend against such intradisk failures, a number of techniques, such as disk scrubbing [3, 28] and checking already written data [30], are being employed. These reliability-improving techniques either periodically examine the contents of the disks for latent sector failures or perform checking reads to ensure that data have been written successfully, so that proactive action can be taken to avoid data loss and exposing the failures to the end user. To minimize performance impact, reliability-improving techniques typically run as background jobs that utilize the periods when the disk is idle to perform data integrity checks. The use of disk idle periods for improving reliability (e.g., for scrubbing), conflicts with energy-saving mechanisms [31–46] that want to turn the disks off during these periods and thus preclude improving reliability. Subsequently, the energy consumption of such systems further increases as depicted in Figure 5.1. This raises the question of whether the energy management and reliability-improving mechanisms can coexist.

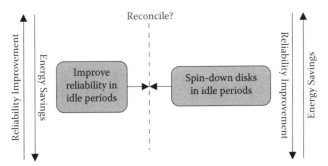

FIGURE 5.1 The impact of improving storage system reliability and energy efficiency on each other.

Another aspect of the issue is that large data centers employ remote-site replication to provide backup in the face of geographically localized disasters. However, such replication is often treated as an add-on and is not taken into consideration in the data center's local storage system design. Therefore, each data center still uses high-reliability systems that consume a significant amount of energy while providing only marginal reliability improvements.

Energy management in large-scale data storage systems is crucial. Energy management in large data storage systems has been receiving extensive research focus [19–21, 33–36, 43–47] as the storage energy consumption increasingly accounts for a higher fraction of the total energy consumption. Given these facts and trends, researchers have started to address disk energy conservation in data-intensive servers [33, 46, 48–51]. One approach is to exploit the fact that server workloads exhibit wide variations in intensity over time. Along these lines, Carrera et al. [48] and Gurumurthi et al. [49] considered multispeed disks. These works showed that significant energy savings can be accrued by adjusting the disk rotational speeds according to the load imposed on the disks. Carrera et al. also showed that a combination of laptop and SCSI (small computer system interface) disks can be even more beneficial in terms of energy, but only for overprovisioned servers. Alternatively, the disk idle times can be increased, so that the disk remains in a low-power mode longer [50, 51]. In contrast, Colarelli and Grunwald [33] proposed the massive array of idle disks (MAID), in which data are copied to "cache disks" to increase idle times at the regular (non-cache) disks. More recently, Pinheiro and Bianchini [46] demonstrated that relying on file popularity and migration, as in their popular data concentration (PDC) technique, produces more robust energy savings than in the MAID approach. Their idea was to migrate the popular data to a subset of the disks, so that other disks would become idle for longer periods. However, their approach may change the reliability behavior of the system by loading some disks more than others, which entails further investigation.

Energy-reliability trade-offs in storage systems have not been explored. Despite the advances in energy management and storage reliability, there has been little research investigating their interaction. A quick literature review revealed a large body of work on either energy management or storage reliability technologies [1–5, 15, 18, 21, 25–29, 31–46, 48–82], but few asked the questions of how reliability improvement techniques impact energy, or conversely, how energy consumption can be reduced while still achieving high reliability.

In this chapter, *we aim to explore the design space of providing energy-efficient and reliable storage systems.* It is clear that both energy efficiency and reliability have to be considered as an integrated approach to provide an energy-efficient and reliable storage system. To help us evaluate the integrated mechanisms that simultaneously optimize energy consumption and reliability, we explore the similarities between energy-performance optimization and energy-reliability optimization to study energy efficiency in the presence of reliability improvement techniques.

Energy management and performance are conflicting optimizations since shutting the disk down will introduce additional delays, potentially degrading the performance. To provide a balanced design, researchers rely on the energy-delay product (*EDP*) [83]. The lowest *EDP* indicates both low energy consumption and low delay, therefore maximizing overall system efficiency. Similarly, as *EDP* is used to capture the effect of energy savings on performance, we present the concept of the energy-reliability product (*ERP*) to capture the combined performance of energy-saving and reliability-improving approaches. This metric can provide a unified mechanism for evaluating both energy efficiency and data reliability in the system. The challenge lies in quantifying reliability and understanding the meaning of *ERP*. There are many possible alternatives (e.g., user-specified expected reliability or energy-savings level). However, we argue that *ERP* is necessary in evaluating different combinations of energy management and reliability scenarios. We need a metric that will allow us to quantitatively compare two different designs, just as *EDP* assists in comparing different designs with respect to energy and performance. We argue that *ERP* is that metric, and we show its applicability using some intuitive examples.

5.3 CURRENT STATE OF THE ART

Extensive research studies have been performed individually on energy management and reliability. However, research effort that takes both areas into consideration has been lacking. In the following, we briefly review the current state of the art in each individual area.

5.3.1 Reliability Improvement Techniques

Failures and errors in storage systems can eventually lead to data loss. To avoid this, all modern systems employ some form of reliability-improving techniques, ranging from simple inter- and intradisk replication to large-scale RAID [25] systems with intradisk redundancy. The goal is to increase mean time to data loss (MTTDL), a metric used for measuring reliability.

5.3.1.1 *Redundant Array of Inexpensive Disks*

RAID [25] stores error-coded data across multiple disks to recover from entire disk failures. In essence, RAID employs some form of error coding, e.g., parity bit, erasure coding, etc., to introduce redundancy. The redundant data is then striped across multiple disks. Depending on the level of redundancy, RAID systems can be designed to avoid single or double disk failures simultaneously (RAID 6). Thus, RAID systems almost always involve an increased number of disks than actually needed to store data to remain active. Consequently, RAID systems will have higher energy consumption than those without it.

5.3.1.2 *Latent Sector Errors*

Disk failure and replacement rates in the field have been found to differ from the mean time to failure (MTTF) specified by drive manufacturers. Schroeder and Gibson [1] analyzed field-gathered disk replacement data from a number of large production systems and showed that accepted techniques of modeling LSEs need refinement to better reflect the realities of how often and at what age disks are replaced. They noted that there is not much difference in replacement rates between different types of disk drives. This indicated a higher influence of operating conditions, rather than that of component-specific factors, on replacement rates. For older systems, specified MTTFs underestimated replacement rates by as much as a factor of 30. The authors also observed increasing failure rates with age rather than a significant infant mortality indicating the effect of early-onset wear and tear. In addition, they found that disk failures were clustered together in space and time on the disk. This showed that multiple LSEs are probably caused by a single event, such as a scratch, rather than by normal wear and tear over a period of time.

These observations motivated the need for refinement in the techniques used to model disk failures. Statistical characteristics of disk failures used to be modeled by a Poisson distribution and hence implied the following:

(1) Disk failures are independent; and (2) the time between failures is exponentially distributed. However, the findings of the mentioned suggested study that failures are not independent.

5.3.1.3 Interleaved Parity Check

Dholakia et al. [26] proposed a new redundancy scheme for high reliability in RAID storage systems. While the implementation of parity in RAID protects against disk failures, the scheme the authors proposed protects against media-related unrecoverable errors. This had become more important in the light of the recent industry trend to use high-capacity but low-reliability drives instead of opting for higher-reliability drives that are more expensive. Their solution was to use a scheme based on interleaved parity check (IPC) where contiguous sectors are arranged logically in a matrix; the sectors in a column are used to obtain the parity sector. The "interleave" comprises the data sectors in the column and the parity sector. Thus, n data sectors combined with $m = interleaving \ depth$ parity sectors form a segment of length l. Dholakia et al. compared the IPC scheme with schemes based on Reed-Solomon codes [84] and single-parity check codes. Their analysis showed that, in case of correlated errors, the IPC scheme provides for high improvement in reliability.

They also compared various redundancy schemes on independent and correlated models of disk failure and compared the reliability of the various schemes. They found that MTTDL scales inversely with the size of the system; that is, for an increase in the system size by a certain factor, MTTDL decreases by the same factor. Analytical and simulation results showed that the performance penalty of using IPC is minimal, thus making a RAID-5 system enhanced with IPC comparable to a RAID-6 system.

5.3.1.4 Disk Scrubbing

With the rise in LSEs in modern high-capacity disks [2], the redundancy provided by RAID can be compromised due to data corruption. This is especially bad in the case of infrequently accessed data, which can become corrupted due to intradisk errors. The errors remain undetected until some disk crashes and RAID fails to recover the data.

To avoid data loss due to intradisk errors, disk scrubbing [4, 28] is employed in conjunction with RAID to discover latent errors in time. Scrubbing is done periodically with a fixed interval between two consecutive cycles, referred to as the scrubbing period. Scrubbing reduces the probability of data loss by reading the whole disk periodically and detecting errors that will not otherwise be found by standard data accesses. A detected error can then be fixed using RAID. Thus, a disk failure in any part of the disk can be discovered as soon as it is scrubbed, and actions can be carried out to recover the corrupted copies from RAID before data loss can happen. The practical scrubbing period—the interval between two consecutive scrubbing cycles—is usually between 1 day and 2 weeks [4].

5.3.1.5 Intradisk Redundancy

In contrast to scrubbing, intradisk redundancy [27] keeps multiple copies of data on a disk to enable recovery from LSEs. This approach does away with periodic scrubbing but may introduce overhead when writing data. Such overheads, however, can be hidden from the end user by appropriately designing the I/O subsystem.

Iliadis and Hu [85] showed that, in light of the fact that the rate at which unrecoverable media errors occur is higher than the manufacturer specifications, there is a need to change the intradisk redundancy schemes. Reliability improvements afforded by intradisk redundancy schemes were found to be adversely affected by the higher number of unrecoverable errors. They proposed revised parameters and showed that the same level of reliability as that of a system without any errors could be achieved with a scheme that incorporates their parameters.

Parameters for intradisk redundancy schemes, such as l and m, are chosen to improve storage efficiency, performance, and reliability. Iliadis and Hu showed that doubling the interleaving depth and the segment length provided the same reliability as that of one without unrecoverable errors. The authors looked into the effect of distribution of length of error bursts. They found that the reliability achieved is dependent on the tail of the distribution, and thus an accurate modeling of the distribution is required. The performance degradation is a modest 2% decrease in saturation throughput for a RAID-5 system enhanced with an intradisk redundancy scheme compared with that of plain RAID-5. Their system proved to be a good choice for achieving high reliability without having to incur the higher expense associated with a RAID-6 array.

5.3.1.6 Idle Read after Write

IRAW (idle read after write) [30] avoids the need for multiple data copies on disks by keeping the data in a disk cache until they has been written and verified. The intuition behind IRAW is that most LSEs occur shortly after writing, and thus by keeping a copy of data in memory for a short period of time, most of the latent errors can be avoided. The challenge is to perform the write verification without affecting I/O performance. This approach can achieve at least as high reliability as scrubbing but may impose performance overheads.

5.3.1.7 Fine-Tuning Intradisk Redundancy and Accelerated Scrubbing

Schroeder et al. [86] have recently provided parameters for a Pareto distribution that fit the data on LSEs very well. The information is leveraged to propose two new intradisk redundancy schemes called MDS+SPC (maximum distance separable erasure encoding scheme + single-parity check) and CDP (column diagonal parity). The authors evaluated their redundancy schemes against many of the aforementioned schemes in isolation, specifically SPC, MDC, and IPC. CDP was found to increase reliability by as much as 30% compared to SPC and achieved higher reliability compared to IPC; however, the latter result could be attributed to different assumptions about the modeling of LSEs. In addition, they also proposed a new scrubbing policy called accelerated scrubbing. The key insight is that LSEs are localized; therefore, the next few sectors (or the rest of the disk) are scrubbed at a higher (accelerated) rate than normal whenever an error is detected. Thus, accelerated scrubbing enabled faster detection of subsequent errors.

Schroeder et al. looked into the length and spacing of error bursts and the spacing between errors and provided parameters for a Pareto distribution fitting these statistical properties. This is a major contribution in that these values can be used by other researchers who want to model LSEs for their work. In addition, they investigated the location of the errors on the drive and found that errors are clearly concentrated on the first part of the drive rather than on the remainder of the drive—"between 20% and 50% of all errors are located in the first 10% of the drive's logical sector space." In addition, there are bumps in the distribution of errors on the disk, indicating the usage pattern might influence the concentration of errors. They also looked at how close in time errors occur and observed that disks experience all LSEs in their lifetime within the same 2-week period.

5.3.1.8 Staggered Scrubbing

Oprea and Juels [87] proposed staggered scrubbing, where the disk is divided into multiple regions, each consisting of multiple segments. The first segment from each region is read in a round-robin fashion, then the second segment and so on. The process starts all over again after the whole disk has been scanned. This approach enabled a faster disk scan. Since errors have been found to be localized, this policy increased the chances of finding one of them quickly. This method does not incur any significant I/O over-head, while higher reliability is achieved without changing the scrubbing frequency.

Oprea and Jules proposed a new metric, mean latent error time (MLET), which measures the "average fraction of the total drive operation time during which it has undetected LSEs." Thus, a lower MLET means the system is less susceptible to data loss. The authors showed that staggered scrubbing improves the mean time to error detection by up to 40%, and MLET by up to 30% compared to fixed-rate sequential strategy. They stressed that it is not possible to come up with a one-size-fits-all scrubbing strategy, and there is a need for adaptive scrubbing strategies that take into account a wider search space.

From this discussion, we observe disk reliability is an area of active research. Various techniques have been proposed over the years to ensure the reliability of disks, but the problem is compounded by the fact that multiple disks are often used together, and that too to ensure reliability against single-disk failures. Thus, the problem of disk reliability is far from solved.

5.3.2 Energy Management Techniques

Energy management in typical enterprises ranges from encouraging users to power down their computers once they leave work to dynamic approaches that detect when devices are not in use and move them into low-power modes. However, after each shutdown, subsequent disk I/Os will require a disk spin-up that consumes energy. For this reason, a *break-even time* is defined as the smallest time interval for which a disk must stay off to justify the extra energy spent in spinning the disk up and down. Also, spin-ups can result in multisecond delays in response and reduce performance and energy efficiency.

The challenge is in identifying idle intervals that would benefit energy efficiency. The most popular approach shuts down the device after a period of inactivity [37, 38], as shown in Figure 5.2. More aggressive dynamic predictor shuts down the device much earlier than the time-out mechanisms have been investigated [39, 40]. To improve accuracy, energy management

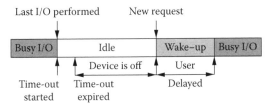

FIGURE 5.2 Anatomy of an idle period.

can be relegated to programmers since they have a better idea of what the application, and potentially the users, are doing at a given time [41, 42]. Finally, power-aware RAID (PARAID) [52] takes into account the power consumption of RAID systems. However, PARAID does not consider intradisk reliability improvement techniques and their energy impact and is orthogonal to the presented work.

5.3.2.1 High-Energy Consumption of Disk Arrays

To store large amounts of data, data servers typically employ large disk arrays. Data from the Transaction Processing Council (TPC) [88] illustrates how large disk arrays may have to be designed for high performance. For example, IBM reports TPC-H [89] results for a server configuration with sixteen 2-GHz AMD Opteron processors and three hundred thirty six 15,000-rpm SCSI disks. Even when the goal is a low price/performance ratio, disk arrays can be large. For example, Hewlett-Packard reports TPC-C [90] results for a server with one 2.4-GHz Intel Xeon processor with thirty five 15,000-rpm SCSI disks.

In these large systems, disk energy consumption can be a serious problem given that high performance is paramount, and high-performance disks consume significant amounts of energy, even when compared to microprocessors, memories, or power supply losses. For instance, in the two examples we mentioned, the energy consumption of the disk array was more than four times higher than that of the processors.

This disk energy problem will worsen in the future, as an increasing number of high-performance disks will be needed to match the performance of each microprocessor. Furthermore, as energy conservation techniques for servers [19–21,47] start to be applied in practice, the disk fraction of the total energy consumption will be even more significant.

A recent paper by Sehgal et al. [91] looked into the behavior of file systems under different workloads. They found, unsurprisingly, significant variations in the amount of work done in unit time and unit energy for different scenarios. Careful tuning of file system parameters yielded significant energy improvements, ranging from 5.4% to 840% over the default configurations. The key insight is that busy servers with predictable workload patterns can take advantage of this fact to ensure careful examination of file system parameters before they start. In addition, the authors experimented with different file systems for different workloads and found that no single file system works well in all cases. Thus, it serves as yet another motivation for looking into the choice of file system and associated parameters.

Another observation reported by Sehgal et al. was a strong linear relationship between the power efficiency and performance of a file system. The authors suggested that the potential power savings from tuning file system parameters is so large that it warrants costly interventions, such as file system reformat, which involves data migration. The authors also reported a guideline based on the results of their experiments for the choice of file system and parameters for different working conditions, for example, using balanced trees as the addressing mechanism when the average file size is in hundreds of kilobytes and using extent-based balanced trees with delayed allocation for files of the order of gigabytes and terabytes.

5.3.2.2 Energy Conservation via Disk Block Migration

Given current trends, researchers have started to address disk energy conservation in data-intensive servers [33, 46, 48–51]. One approach exploits the fact that server workloads exhibit wide variation in I/O intensity, which in turn can be leveraged to conserve energy. Along these lines, Carrera et al. [48] and Gurumurthi et al. [49] considered multispeed disks. These papers showed that significant energy savings could be accrued by adjusting the disk rotational speeds according to the load imposed on the disks. Carrera et al. also showed that a combination of laptop and SCSI disks can be even more beneficial in terms of energy, but only for overprovisioned servers.

Another approach increases disk idle times, so that disks can be sent to low-power modes and be kept there for longer periods of time. Zhu et al. [50] considered storage cache replacement techniques that selectively keep blocks from certain disks in the main memory cache to increase their idle times. Recently, Zhu et al. [51] studied a more elegant energy-aware storage cache replacement policy in which dynamically adjusted memory partitions were used for caching data from different disks. These studies did not involve data movement, which could provide further energy savings.

In contrast, Colarelli and Grunwald [33] proposed MAID, in which data are copied to "cache disks" to increase idle times at the regular (noncache) disks. More recently, Pinheiro and Bianchini [46] demonstrated that relying on file popularity and migration, as in their PDC technique, produces more robust energy savings than in the MAID approach. Their idea was to migrate the popular data to a subset of the disks, so that other disks would become idle for longer periods. However, their approach has two limitations: (1) It is based on file access behavior; and (2) it may take a relatively long time to categorize popular data as such.

Given these limitations, their system would not work well for servers that do not access entire files at a time. In particular, database servers access data with block granularity and exhibit widely different popularities for different blocks. This paper addresses these limitations by using a program counter (PC)-based approach to determine the popularity of disk blocks and migrates them around to offload some of the disks.

Verma et al. [92] chose a few volumes from each disk that constituted its working volume, and only the working volume was replicated. This resulted in a storage virtualization layer optimization that did not involve expensive data migration across physical volumes. The key insight is that the working set in a storage system constitutes a small and stable workload that can be efficiently replicated. Thus, all disk accesses can be effectively routed to a small number of disks, while the other disks and volumes can remain idle.

The authors described proportional power savings of around 35% by using their solution in a prototype experimental setup. They achieved higher savings of around 59% using an aggressive estimate of disk I/O per second (IOPS).

5.3.2.3 Energy Conservation in Remote Replication Scenarios

Replication of data in data centers ensures reliability; however, to provide disaster recovery guarantees, they need to be replicated in a remote location. In addition to the cost of running a remote data center, significant costs are incurred during communication between the two. Goda and Kitsure-gawa [93] looked into the problem of how to achieve power-aware remote replication.

The key insight here is that remote centers often serve as a store for various copies of data rather than as a primary storage place. Hence, small penalties are acceptable if the energy savings are large enough. The changes to the primary storage are transferred as logs to the remote center, and it applies the logs to the replicas, similar to databases. The authors achieved energy savings by delaying the application of the transferred database log. This allows the disks used to store the data volumes in the remote site to remain idle for a longer duration, thereby leading to significant energy savings. In addition, the authors used eager log compaction to reduce the amount of log that needs to be applied and thus improve log application throughput. Log compaction is achieved by log folding, a strategy of coalescing a series of log updates in a window of time to a single log update indicating the final result of the series of updates. They also sorted the logs in the window to improve the sequentiality of disk accesses. They were able to do this becase database logs store physical references to target records. With small penalties of possible service breakdown time, the authors managed to achieve 80–85% energy savings on the remote center.

5.3.2.4 Energy Conservation via Novel Storage Technologies

A growing trend in designing storage systems is introduction of solid-state storage devices [58, 94–97], which can provide both performance and energy benefits. However, solid-state disks that are constructed using flash memory have several limitations that can limit performance, lifetime, and energy efficiency of the device. An erase before update combined with random small-size writes can significantly degrade performance resulting in performance level below mechanical drives. A limited number of writes to flash storage can limit the lifetime of the device. In addition, I/O-intensive workloads may encounter device failures after exhausting its maximum number of write-erase cycles. While individual devices are more energy efficient than mechanical hard drives, the lower capacity of the solid-state drives requires several devices just to match the capacity of one mechanical drive. As a result, the energy benefits may be unclear and require a detailed study. Finally, the simple spinning-disk-based reliability-energy model may not hold for such devices; thus, solid-state storage requires a detailed study to identify the key trade-offs between energy efficiency and reliability.

One such recent work was by Andersen et al. [98], who coupled low-power embedded CPUs to local flash storage. This setup allows for providing fast and efficient yet low-power access to a large, random-access storage system that serves accesses to predominantly small files, as in a key-value lookup system. In fact, the authors designed a key-value store using their setup to show that it is practical to use embedded flash-memory-enabled CPUs. The authors observed that CPU power consumption grows super-linearly with speed on one hand; on the other hand, dynamic power scaling on traditional systems is inefficient. In such a scenario, their choice of low-power embedded CPUs provides the best of both worlds. Flash consumes much less power than traditional disk drives, is a lot cheaper than comparable DRAM (dynamic random access memory), and is particularly useful for random access because it does not incur disk-spinning delay. Thus, coupling them provided the authors with a best-case choice of components to construct a key-value store.

However, the authors conceded that their design of the key-value store was tailored to the peculiarities of a flash-based system. Thus, while flash has been proven to be a viable low-cost alternative to traditional disks, widespread adoption of the same is still a challenge.

Yet another novel architecture has been proposed by Zhu et al. [15], who utilized multiple disks that at different speeds to reduce the overall energy consumption. The intuition here is that different data sets are accessed at different frequencies. Thus, by keeping data with similar access patterns on one disk, only the currently accessed disk needs to stay on. Other disks can remain in standby state. Their technique, called hibernator, uses a coarse-grained algorithm to determine the workload. It transparently determines the speed settings for the disks and migrates data to the disks with the proper speeds. To ensure service guarantees are met, it can boost the disk speeds if performance goals are at risk. This approach provides for energy savings by creating opportunities for disks to stay off, while ensuring that the data being accessed can be serviced without performance degradation.

The authors evaluated their technique using trace-driven simulations with traces of real file systems and online transaction processing (OLTP) workloads. They also used a hybrid system comprised of a real database server and a storage server. They compared hibernator against various other previously proposed schemes for disk energy management. The authors found that hibernator provided significant energy savings of around 65%, which is 6.5 to 26 times better than previous solutions.

5.4 *ERP:* BALANCING RELIABILITY AND ENERGY EFFICIENCY

We face two alternatives for utilizing disk idle time. On one hand, the idle period provides opportunities for spinning down the disk and saving energy; on the other hand, it can provide time for techniques to improve reliability without performance penalties. While advanced techniques exist for both, we initially focus on how basic energy savings (e.g., time-out based

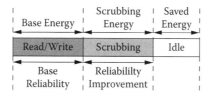

FIGURE 5.3   Disk activity in the system.

shutdown) and reliability-improving approaches (e.g., disk scrubbing) (Section 5.3), can coexist. The goal is to evaluate basic mechanisms for idle time allocation between energy management and reliability mechanisms to achieve a balance between energy consumption and reliability.

Figure 5.3 shows an example of a time distribution for some workload. The time spent in performing I/O requests is independent of the scrubbing or energy-saving mechanisms. Energy consumed for satisfying I/O requests is dictated by the application behavior in the system. During busy periods, the reliability of the disk drive is not affected and corresponds to the basic disk reliability [27]. However, when the disk is idle, part of the idle time can be used to improve reliability and part to save energy. The key observation here is that if more of the idle time is allotted to disk scrubbing, the expected MTTDL will increase, which in turn decreases the probability of failures and thus improves reliability. However, less time will be available for spinning the disk down, and additional energy will be consumed to perform scrubbing, thus decreasing energy efficiency. Spending more energy will increase reliability; however, the goal is to minimize energy consumption while maximizing reliability.

5.4.1   Efficiency and Reliability Metrics

System designers are concerned with both power and energy when designing large data centers. Power levels of the running machines dictate the cooling and energy delivery infrastructure. Energy associated with the execution of the application directly translates into monetary costs required to purchase a given amount of energy. However, reducing the power and energy usage of an application usually translates to degradation in performance. Therefore, system researchers have always been concerned in evaluating the energy versus performance trade-off. The trade-off originated in circuit design since CMOS (complementary metal-oxide semiconductor) circuits are able to reduce energy consumption at a cost of performance. Subsequently, computer architects proposed the *EDP*, as a natural metric to simultaneously capture the impact of energy and performance [83] and, in turn, characterize the energy-performance design space with the intent to identify the design with the best combination of energy and performance efficiency.

EDP was further augmented into an energy delay squared product, ED^2P, that was argued to be independent of voltage in evaluating energy efficiency of dynamic voltage and frequency scaling of processors [99]. However, we have to be careful in applying *EDP* or ED^2P to interval-based energy efficiency mechanisms. While each interval may result in optimal energy efficiency, the overall application execution may not [100]. Therefore, it is critical to carefully consider each metric and explore the effect on overall energy efficiency. Similarly, on designing an *ERP*, it is critical to explore the design issues before applying it carelessly to every situation.

5.4.1.1   Choice of ERP Metric

We can draw an analogy to energy-performance optimizations, where energy savings are increased at a cost of decrease in performance. The popular metric to combine both performance impact and energy savings is expressed as an *EDP*. Consequently, researchers are faced with the task of minimizing a single value, the corresponding *EDP* [83], to improve overall system efficiency. To provide a similar metric for evaluating energy and reliability, we propose the *ERP* to capture the impact of energy and reliability improving techniques on each other. We focus on the amount of additional energy consumed due to scrubbing and the increase in reliability it generates. We do not include the base energy consumption or reliability since it is independent of energy management or reliability mechanisms. The *ERP* can also be expressed in terms of energy savings generated in the system and the increase in reliability due to scrubbing.

Since reliability cannot be simply measured, we define the *ERP* in terms of MTTDL as follows:

$$
\begin{aligned}
ERP &= \textit{Energy savings} * \textit{Reliability improvement} \\
&= \Delta \textit{Energy} * \Delta \textit{MTTDL},
\end{aligned} \tag{5.1}
$$

where $\Delta \textit{Energy}$ is the energy saved, and $\Delta \textit{MTTDL}$ is the corresponding improvement in reliability. Recently, detailed models [26, 27] have been developed that show how MTTDL is affected by varying scrubbing period

TABLE 5.1 Western Digital
WD2500JD Specifications

State	
Read/write power	10.6 W
Seek power	13.25 W
Idle power	10 W
Standby power	1.8 W
Spin-up energy	148.5 J
Shutdown energy	6.4 J
State Transition	
Spin-up time	9 s
Shutdown time	4 s

when disk scrubbing is employed. Based on these models, under typical workload and scrubbing periods, the MTTDL increases with decreasing scrubbing period. Thus, Equation 5.1 can be simplified to

$$\Rightarrow ERP \propto \Delta Energy * 1/Scrubbing\ Period \qquad (5.2)$$

This leaves us with the objective to maximize *ERP* through idle time allocation that achieves the best combination of energy savings and reliability.

5.5 APPLYING *ERP*

In this section, we discuss the evaluation of *ERP* as a suitable metric for capturing the energy and reliability constraints. In the following, we describe the experimental setup and present our evaluation using typical desktop applications.

5.5.1 Methodology

Table 5.1 shows the specifications of the disk that we used in our simulation. The Western Digital WD2500JD disk has a spin-up time of about 9 s from a sleep state, which is common in high-speed commodity disks. Similarly, the other numbers mentioned are also typical for such disks.

Table 5.2 shows six desktop applications that are popular in enterprise environments: *Mozilla* web browser, *Mplayer* music player, *Impress* presentation software, *Writer* word processor, *Calc* spreadsheet, and *Xemacs* text editor. These applications were chosen because they have been used in previous studies and are well studied [31, 82]. Mozilla is a web browser with

TABLE 5.2   Traces Collected for the Studied Applications

Application	Trace Length (h)	Number of Reads	Writes	Referenced (MB) Reads	Writes
Mozilla	45.97	13,005	2,483	66.4	19.4
Mplayer	3.03	7,980	0	32.3	0
Impress	66.76	13,907	1,453	92.5	40.1
Writer	54.19	7,019	137	43.8	1.2
Calc	53.93	5,907	93	36.2	0.4
Xemacs	92.04	23,404	1,062	162.8	9.4

which a user spends time reading page content and following links. In this case, I/O behavior depends on the content of the page and user behavior. Impress, Writer, and Calc are part of the Open Office suite of applications, and all three are interactive applications with both user-driven I/O behavior and periodic automated I/O behavior (i.e., autosaves). Writer is a word processor used to compose documents and perform quick proofreading corrections. Impress is intended for creation and editing of presentations. Calc is a spreadsheet. Xemacs is primarily used to edit one or more text files of various size. Finally, Mplayer is a media player that is generally active only when a user watches a media clip. Detailed traces of user-interactive sessions for each application were obtained by a *strace*-based tracing tool [101] over a number of days. Finally, the table also shows trace length and the details of I/O activity. I/Os satisfied by the buffer cache were not counted since they do not cause disk activity.

5.5.2   Evaluation

To better illustrate the role of *ERP*, Figure 5.4 presents corresponding energy savings as well as improvement in reliability for two of the applications [24]. For each of the studied applications, the energy consumed and the time it would take for a scrubbing cycle to complete are simulated. Next, the scrubbing period to estimate the improvement in reliability is used, and finally Equation 5.2 is used to determine the *ERP* for each measurement. For each application, an increasing portion of an idle period is dedicated to scrubbing (0% to 100%), and the resulting changes in reliability and energy as well as the *ERP* are plotted. The reason for using such simple mechanisms is to understand the meaning of the value conveyed by *ERP*. This allows us to reason about the impact of *ERP* and to observe whether *ERP* follows the intuition about energy savings and reliability improvements.

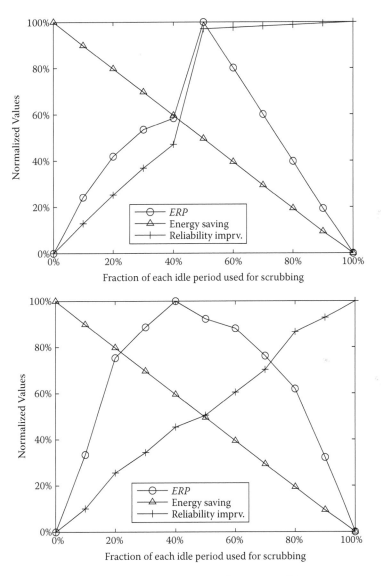

FIGURE 5.4 Normalized energy savings and associated improvements in reliability under varying fractions of idle periods. *(continued)*

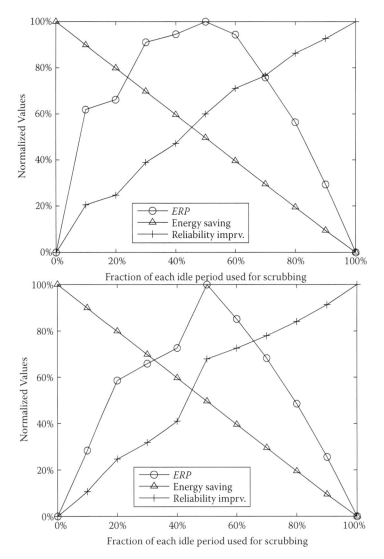

FIGURE 5.4 *(continued)*. Normalized energy savings and associated improvements in reliability under varying fractions of idle periods. *(continued)*

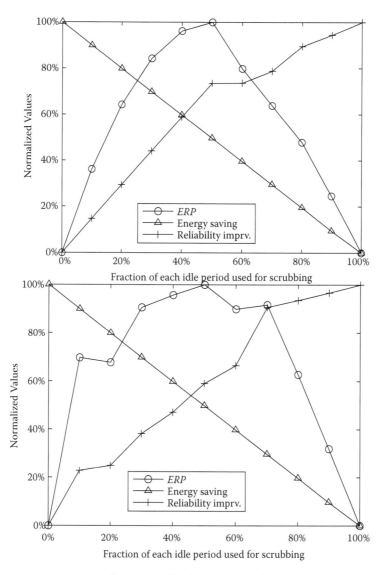

FIGURE 5.4 *(continued)*. Normalized energy savings and associated improvements in reliability under varying fractions of idle periods.

Figure 5.4 also includes an *ERP* of the mechanism that tries to balance both energy and reliability by utilizing all idle periods that are smaller than the break-even time for scrubbing. This does not affect the energy savings but improves reliability. Observe that the peak *ERP* points to a distribution of idle periods that can provide an efficient combination of energy savings and reliability. These results show that *ERP* is capable of capturing the trade-off point where energy savings can be realized without impacting reliability and vice versa.

Next, the authors looked [24] into various alternate schemes for allocating idle periods. In the first scheme, they adopted a two-phase approach. In the first phase, all idle periods are dedicated for scrubbing. This phase continues until the entire disk is scrubbed once. Then, the second phase starts; all idle periods are used exclusively for energy savings, and no scrubbing is done. In this case, the scrubbing period equals the trace length.

The "Two-Phase" column of Table 5.3 shows the energy consumed and the associated reliability improvement for each of the studied applications. The numbers are normalized against those of the previous experiments. The results show that although this scheme saves energy, the reliability improvement is very small, with an *ERP* of less than 24.8% averaged across all applications. This approach is not advisable for two reasons: (1) the scrubbing cycle is significantly large and the resulting MTTDL is very small; and (2) in the energy-saving phase, all idle periods that are smaller than the break-even time are wasted, as they neither provide energy savings nor contribute to improving reliability.

In the second scheme, periods smaller than break-even time are used for scrubbing—avoiding loss in energy savings—and longer periods for

TABLE 5.3 Energy Savings and Improvement in Reliability Achieved under Different Idle-Period Allocations

Application	Two-Phase Allocation (%)		Scrubbing Only in Small Idle Periods (%)		Alternate Allocation (%)	
	Energy Savings	Reliability Improvement	Energy Savings	Reliability Improvement	Energy Savings	Reliability Improvement
Mozilla	96.2	5.1	99.9	0.7	54.3	35.5
Mplayer	42.6	84.4	99.9	0.0	99.9	0.0
Impress	97.5	2.6	99.9	0.4	41.6	71.3
Writer	96.8	3.2	99.9	0.3	46.8	60.9
Calc	96.8	4.1	99.9	0.6	30.2	43.8
Xemacs	98.1	2.4	99.9	0.1	70.9	74.5

spinning off the disks. Table 5.3 also shows the result for this approach. Once again, although energy savings are maximized ($\approx$100%), the scrubbing period increases significantly, thus resulting in negligible ($\leq$ 0.7%) reliability improvement and an $ERP \approx$ 0%.

In the third scheme, scrubbing the disk and spinning down the disk are alternated on each idle period. All idle periods smaller than the break-even time are once again used only for scrubbing. The "Alternate" column of Table 5.3 shows the results. Overall, this scheme achieves good energy savings (average 47.7%) while improving reliability (average 57.3%). The average ERP of 76.4% (179.5% for Xemacs) suggests that this is a viable approach compared to the first two.

Finally, the reliability impact of the commonly used time-out-based approach for energy management was also investigated. A timer with a fixed time-out interval is started whenever the disk is idle, and the disk is spun down when the timer expires. Scrubbing is only done during the time-out interval, that is, the time between the disk becoming idle and being shut down.

Figure 5.5 shows the resulting energy savings and reliability improvements for the studied applications under various time-out intervals. Note that for this case, the energy comparison is done against a simple time-out scheme without scrubbing. It is observed that such an approach, while providing almost 100% of the possible energy savings, does not provide enough opportunity for scrubbing, thus achieving little reliability improvement. The problem with simply using a longer time-out is that the resulting reliability improvement depends on the idle period distribution. For example, if there are a large number of idle periods, the longer time-out would provide desired higher reliability at a cost of energy savings. On the other hand, if we imagine a scenario where there is only one very long busy period followed by one long idle period, the one time-out interval spent on scrubbing is not sufficient for significantly improving reliability.

5.6 EXTENDING *ERP*

In an initial exploration [24], it has been shown that the *ERP* function can identify trade-off points that provide for improving disk reliability with minimal impact on energy savings. Nonetheless, it is clear that further study is needed to broadly understand the implication of *ERP* and its impact on designing both energy-efficient and reliable systems. Once the implications are clear, further research is needed in more dynamic energy-reliability

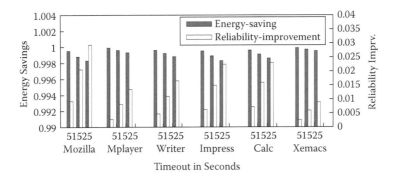

FIGURE 5.5 Normalized energy savings (gray) and improvement in reliability (white) achieved under a time-out-based approach with different time-outs.

management techniques. Different workload characteristics of the applications affect the portion of time dedicated to one of either the energy or reliability improvement techniques. Therefore, adaptive techniques are needed that dynamically monitor system characteristics and provide online energy-reliability balancing mechanisms. The analysis and identification of such use cases remain a focus of ongoing research.

Another problem with such analysis is the relative weight of the reliability and energy in a unification metric. Should energy be a harder constraint thus justifying an E^2RP or vice versa an ER^2P metric for reliability? Moreover, the energy-reliability relationship is explored in terms of distribution of disk idle times. A more traditional relationship between the two also exists; that is, more power dissipation can lead to thermal stress on the disk, causing it to fail. For now, it has been assumed that the systems are properly cooled and remain within acceptable working thermal limits; thus, we do not capture this. However, the impact of such factors also needs to be explored.

Alternatives to scrubbing, such as intradisk redundancy [27] and IRAW [30], require modifying the reliability-measuring approach discussed so far. Finally, the simple disk model that has been used so far can be extended to handle advanced multispeed disks or sophisticated controls such as varying disk speeds [49]. Although a single disk with multiple speeds is currently impractical, an advanced model may help capture disk arrays with heterogeneous disks. In addition, while this work focuses on disks, it can be extended to other components within a computing system.

Nontraditional temperature range operation: Recent trends suggest that the system components can be pushed to operate beyond suggested temperature ranges, affecting reliability but ultimately saving energy that otherwise would be spent on cooling. While such trends are already becoming a reality, the exact trade-offs are unknown. There is a need to explore the relationships between component costs, their reliability, and potential energy savings that result in reducing component reliability. One potential goal is to minimize the overall cost of doing business, while the other is to look at the energy savings versus reliability, the *ERP*. While costs are important, they may be affected by many other issues and change daily. The *ERP* presents a more stable metric that, once optimized, would result in a clearer scenario for decisions about system operation and maintenance.

Solid-state and mechanical disk hybrid setups: A growing trend in designing storage systems is the introduction of solid-state storage devices [58, 94–97], which can provide performance benefits. The simple spinning-disk-based reliability-energy model may not hold for such devices; thus, a hybrid setup with both solid-state and mechanical storage has to be studied in detail to identify the key trade-offs.

Disk clusters: Large storage systems, such as those employed in modern data centers, often comprise hundreds to thousands of disks. Thus, even with very reliable disks, some disks are always failing in such systems. For instance, a disk crash every 5 to 10 min is a common occurrence. A well-established strategy to avoid data loss in face of such crashes is RAID [25]; a combination of redundant disks and error coding of data are employed to withstand multiple disk failures [102]. As discussed, LSEs are quickly becoming a predominant cause of data loss in modern disks. To guard against such errors, a number of techniques have been proposed to detect the errors early and to avoid data loss by performing active recovery using redundancy available through RAID. As discussed, scrubbing [3, 28] and IRAW [30] are commonly employed for this purpose. It has been argued [27] that intradisk redundancy is better than disk scrubbing against data loss. We aim to study how these techniques impact energy consumption in storage systems.

Reproducible data: Stored data can be classified into two categories: (1) nonreproducible data, which once lost, cannot be (easily) reconstructed, such as, observation data from experiments such as LHC [103]; and (2) reproducible data, which can be regenerated if needed. We argue that while nonreproducible data have stringent reliability requirements, reproducible data offer interesting trade-offs. For example, if the cost of ensuring reliability of a data item exceeds that of reproducing it, then a better option would be to reproduce the data on loss and vice versa. *ERP* can help in investigating the trade-offs provided by the utilization of geographically distributed data redundancy in improving energy and data reliability at the local data center.

The relationship between *ERP* and the monetary costs should be understood, and we hope further research into *ERP* will illuminate this relationship. The metric potentially will provide data center planners with a tool to maximize the utility, while existing data center managers can reevaluate their current operating procedures to improve center efficiency.

5.7 SUMMARY

In this chapter, we explored the interactions between reliability and energy management in disks. We argued that similar to the way the EDP metric is used for capturing the effect of energy management on system performance, we should consider an *ERP* to evaluate the combination of reliability and energy efficiency of storage systems. Using trace-driven simulations of several enterprise applications, we have shown that *ERP* can help identify efficient distribution of disk idle time to energy and reliability management. In this study, we relied on simpler techniques for both energy and reliability management, but the evaluation techniques using *ERP* can guide the design of more advanced energy-reliability management.

REFERENCES

1. Bianca Schroeder and Garth A. Gibson. Disk failures in the real world: What does an MTTF of 1,000,000 hours mean to you? In *FAST*, February 14–16, 2007 San Jose, CA.
2. Eduardo Pinheiro, Wolf-Dietrich Weber, and Luiz André Barroso. Failure trends in a large disk drive population. In *FAST*, February 14–16, 2007 San Jose, CA.
3. Hannu H. Kari. *Latent Sector Faults and Reliability of Disk Arrays*. PhD thesis, Helsinki University of Technology, 1997.
4. Lakshmi N. Bairavasundaram, Garth R. Goodson, Shankar Pasupathy, and Jiri Schindler. An analysis of latent sector errors in disk drives. In *SIGMETRICS*, June 2007 San Diego, CA.
5. Lakshmi N. Bairavasundaram, Andrea C. Arpaci-Dusseau, Remzi H. Arpaci-Dusseau, Garth R. Goodson, and Bianca Schroeder. An analysis of data corruption in the storage stack. *Trans. Storage*, 4(3):1–28, 2008.
6. Google Analytics loses a weeks e-commerce data. http://www.blogstorm.co.uk/google-analytics-loses-a-weeks-e-commerce-data/, 2008.
7. Some Gmail accounts were cleaned out. http://blogs.zdnet.com/Google/?p=432, 2006.
8. Data loss at Google Reader. http://www.techcrunch.com/2007/06/11/data-loss-at-google-reader/, 2007.

9. Facebook data loss fiasco. http://www.sophos.com/blogs/gc/g/2008/12/01/facebook-data-loss-fiasco/, 2008.

10. Jeffrey Dean and Sanjay Ghemawat. MapReduce: Simplified data processing on large clusters. *Comm. ACM*, 51(1):107–113, 2008.

11. Hadoop: Open source implementation of MapReduce. http://lucene.apache.org/hadoop/, 2009.

12. Michael Isard, Mihai Budiu, Yuan Yu, Andrew Birrell, and Dennis Fetterly. Dryad: Distributed data-parallel programs from sequential building blocks. In *EuroSys*, March 2007 Lisbon, Portugal.

13. Michael Isard, Vijayan Prabhakaran, Jon Currey, Udi Wieder, Kunal Talwar, and Andrew Goldberg. Quincy: Fair scheduling for distributed computing clusters. In *SOSP*, October 2009 Big Sky, MT.

14. Christopher Olston, Benjamin Reed, Utkarsh Srivastava, Ravi Kumar, and Andrew Tomkins. Pig Latin: A not-so-foreign language for data processing. In *SIGMOD*, June 2008 Vancouver, Canada.

15. Qingbo Zhu, Zhifeng Chen, Lin Tan, Yuanyuan Zhou, Kimberly Keeton, and John Wilkes. Hibernator: Helping disk arrays sleep through the winter. In *SOSP*, October 2005 Brighton, England.

16. Partha Ranganathan. A science of power management. Invited talk at NSF workshop on the science of power management, April 2009.

17. Xiaodong Li, Zhenmin Li, Pin Zhou, Yuanyuan Zhou, Sarita V. Adve, and Sanjeev Kumar. Performance-directed energy management for storage systems. *IEEE Micro*, 24(6):38–49, 2004.

18. Eduardo Pinheiro, Ricardo Bianchini, and Cezary Dubnicki. Exploiting redundancy to conserve energy in storage systems. *SIGMETRICS Perform. Eval. Rev.*, 34(1):15–26, 2006.

19. Jeffrey S. Chase, Darrell C. Anderson, Prachi N. Thackar, Amin M. Vahdat, and Ronald P. Boyle. Managing energy and server resources in hosting centers. In *SOSP*, October 2001 Banff, Canada.

20. Mootaz Elnozahy, Michael Kistler, and Ramakrishnan Rajamony. Energy conservation policies for web servers. In *USITS*, March 2003 Seattle, WA.

21. Eduardo Pinheiro, Ricardo Bianchini, Enrique V. Carrera, and Taliver Heath. Load balancing and unbalancing for power and performance in cluster-based systems. In *Workshop on Compilers and Operating Systems for Low Power*, September 2001 Barcelona, Spain.

22. NSF science of power management workshop. http://scipm.cs.vt.edu.

23. Daniel A. Reed. Invited talk at NSF workshop on the science of power management, April 2009 Arlington, VA.

24. Guanying Wang, Ali R. Butt, and Chris Gniady. On the impact of disk scrubbing on energy savings. In *USENIX OSDI HotPower Workshop*, December 2008 San Diego, CA.

25. David A. Patterson, Garth Gibson, and Randy H. Katz. A case for redundant arrays of inexpensive disks (RAID). In *SIGMOD*, 1988.

26. Ajay Dholakia, Evangelos Eleftheriou, Xiao-Yu Hu, Ilias Iliadis, Jai Menon, and K. K. Rao. A new intra-disk redundancy scheme for high-reliability

RAID storage systems in the presence of unrecoverable errors. *Trans. Storage*, 4(1):1–42, 2008.

27. Ilias Iliadis, Robert Haas, Xiao-Yu Hu, and Evangelos Eleftheriou. Disk scrubbing versus intra-disk redundancy for high-reliability RAID storage systems. In *SIGMETRICS*, October 2008 Annapolis, MD.

28. Thomas J. E. Schwarz, Qin Xin, Ethan L. Miller, Darrell D. E. Long, Andy Hospodor, and Spencer Ng. Disk scrubbing in large archival storage systems. In *MASCOTS*, October 2004 Volendam, The Netherlands.

29. Weihang Jiang, Chongfeng Hu, Yuanyuan Zhou, and Arkady Kanevsky. Are disks the dominant contributor for storage failures? A comprehensive study of storage subsystem failure characteristics. *Trans. Storage*, 4(3):1–25, 2008.

30. Alma Riska and Erik Riedel. Idle read after write: IRAW. In *ATC*, June 2008 Boston, MA.

31. Chris Gniady, Ali R. Butt, Y. Charlie Hu, and Yung-Hsiang Lu. Program counter-based prediction techniques for dynamic power management. *Trans. Computers*, 55(6):641–658, 2006.

32. Yung-Hsiang Lu, Eui-Young Chung, Tajana Simunic, Luca Benini, and Giovanni De Micheli. Quantitative comparison of power management algorithms. In *DATE*, 2000.

33. Dennis Colarelli and Dirk Grunwald. Massive arrays of idle disks for storage archives. In *Supercomputing*, November 2002 Chicago, IL.

34. Nagapramod Mandagere, Jim Diehl, and David Du. GreenStor: Application-aided energy-efficient storage. In *MSST*, September 2007 San Diego, CA.

35. Mark W. Storer, Kevin M. Greenan, Ethan L. Miller, and Kaladhar Voruganti. Pergamum: Replacing tape with energy efficient, reliable, disk-based archival storage. In *FAST*, February 2008 San Jose, CA.

36. Peter Bodik, Michael Paul Armbrust, Kevin Canini, Armando Fox, Michael Jordan, and David A. Patterson. *A case for adaptive datacenters to conserve energy and improve reliability*. Technical report, EECS Department, University of California, Berkeley, 2008.

37. Fred Douglis, Padmanabhan Krishnan, and Brian Bershad. Adaptive disk spin-down policies for mobile computers. In *USENIX Symposium on Mobile and Location-Independent Computing*, 1995.

38. David P. Helmbold, Darrell D. E. Long, and Bruce Sherrod. A dynamic disk spin-down technique for mobile computing. In *Mobile Computing and Networking*, 38 pages 130–142, November 1996.

39. Eui-Young Chung, Luca Benini, and Giovanni De Micheli. Dynamic power management using adaptive learning tree. In *ICCAD*, November 1999 San Jose, CA.

40. Chi-Hong Hwang and Allen CH Wu. A predictive system shutdown method for energy saving of event driven computation. *ACM Trans. Design Automation Electron. Syst.*, 5(2):226–241, 2000.

41. Andreas Weissel, Bjoern Beutel, and Frank Bellosa. Cooperative I/O—a novel I/O semantics for energy-aware applications. In *OSDI*, October 2002 Hollywood, CA.

42. Carla Schlatter Ellis. The case for higher-level power management. In *HOTOS*, page 162, 1999.

43. Hogil Kim, Yuho Jin, and Eun Jung Kim. *Power management in RAID server disk system using multiple idle states*. Technical report, TAMU Computer Science and Engineering, 2004 College Station, TX.

44. Xiaodong Li, Zhenmin Li, Yuanyuan Zhou, and Sarita Adve. Performance directed energy management for main memory and disks. *Trans. Storage*, 1(3):346–380, 2005.

45. Timothy Bisson, Scott A. Brandt, and Darrell D.E. Long. A hybrid disk-aware spin-down algorithm with I/O subsystem support. In *IPCCC*.

46. Eduardo Pinheiro and Ricardo Bianchini. Energy conservation techniques for disk array-based servers. In *ICS*, June 2004 Saint-Melo, France.

47. Ricardo Bianchini and Ramakrishnan Rajamony. *Power and energy management for server systems*. Technical report, Department of Computer Science, Rutgers University, New Brunswick, NJ, 2003.

48. Enrique V. Carrera, Eduardo Pinheiro, and Ricardo Bianchini. Conserving disk energy in network servers. In *ICS*, June 2003 San Francisco, CA.

49. Sudhanva Gurumurthi, Anand Sivasubramaniam, Mahmut Kandemir, and Hubertus Franke. DRPM: Dynamic speed control for power management in server class disks. *SIGARCH Comp. Arch. News*, 31(2):169–181, 2003.

50. Qingbo Zhu, Francis M. David, Christo F. Devaraj, Zhenmin Li, Yuanyuan Zhou, and Pei Cao. Reducing energy consumption of disk storage using power-aware cache management. In *HPCA*, June 2004 Madrid, Spain.

51. Qingbo Zhu, Asim Shankar, and Yuanyuan Zhou. PB-LRU: A self-tuning power aware storage cache replacement algorithm for conserving disk energy. In *ICS*, June 2004 Saint-Melo, France.

52. Charles Weddle, Mathew Oldham, Jin Qian, An-I Andy Wang, Peter Reiher, and Geoff Kuenning. PARAID: A gear-shifting power-aware RAID. In *FAST*, February 2007 San Jose, CA.

53. Dong Li and Jun Wang. EERAID: Energy efficient redundant and inexpensive disk array. In *ACM EW11*, 2004.

54. Xiaoyu Yao and Jun Wang. RIMAC: a novel redundancy-based hierarchical cache architecture for energy efficient, high performance storage systems. *SIGOPS Oper. Syst. Rev.*, 40(4):249–262, 2006.

55. Kevin M. Greenan, Darrell D. E. Long, Ethan L. Miller, Thomas J. E. Schwarz, and Jay J. Wylie. A spin-up saved is energy earned: Achieving power-efficient, erasure-coded storage. In *HotDep*, 2008.

56. Hyo J. Lee, Kyu H. Lee, and Sam H. Noh. Augmenting RAID with an SSD for energy relief. In *HotPower*, December 2008 San Diego, CA.

57. Euiseong Seo, Seon-Yeong Park, and Bhuvan Urgaonkar. Empirical analysis on energy efficiency of Flash-based SSDs. In *HotPower*, December 2008 San Diego, CA.

58. Dushyanth Narayanan, Eno Thereska, Austin Donnelly, Sameh Elnikety, and Antony Rowstron. Migrating server storage to SSDs: Analysis of trade-offs. In *EuroSys*, March 2009 Nuremberg, Germany.

59. Yan Zhang, Sudhanva Gurumurthi, and Mircea R Stan. SODA: Sensitivity based optimization of disk architecture. In *DAC Conference*, June 2007 San Francisco, CA.

60. Sriram Sankar, Sudhanva Gurumurthi, and Mircea R. Stan. Sensitivity based power management of enterprise storage systems. In *MASCOTS*, September 2008 Baltimore, MD.

61. Sriram Sankar, Sudhanva Gurumurthi, and Mircea R. Stan. Intra-disk parallelism: An idea whose time has come. *SIGARCH Comp. Arch. News*, 36(3):303–314, 2008.

62. Nikolai Joukov and Josef Sipek. GreenFS: Making enterprise computers greener by protecting them better. In *Eurosys*, pages 69–80, March 2008 Glasgow, Scotland.

63. Edmund B. Nightingale and Jason Flinn. Energy-efficiency and storage flexibility in the Blue file system. In *OSDI*, December 2004 San Francisco, CA.

64. RuGang Xu. Conquest: Combining battery-backed RAM and threshold-based storage scheme to conserve power. *WIP Report in SOSP 2003*, 2003.

65. Hai Huang, Wanda Hung, and Kang G. Shin. FS2: Dynamic data replication in free disk space for improving disk performance and energy consumption. *SIGOPS Oper. Syst. Rev.*, 39(5):263–276, 2005.

66. Lanyue Lu and Peter Varman. Workload decomposition for power efficient storage systems. In *HotPower*, December 2008 San Diego, CA.

67. Dushyanth Narayanan, Austin Donnelly, and Antony Rowstron. Write off-loading: Practical power management for enterprise storage. *Trans. Storage*, 4(3):1–23, 2008.

68. Dushyanth Narayanan, Austin Donnelly, Eno Thereska, Sameh Elnikety, and Antony Rowstron. Everest: Scaling down peak loads through I/O off-loading. In *OSDI*, December 2008 San Diego, CA.

69. Le Cai and Yung-Hsiang Lu. Power reduction of multiple disks using dynamic cache resizing and speed control. In *ISLPED*, October 2006 Tegernsee, Germany.

70. Feng Chen and Xiaodong Zhang. Caching for bursts (C-Burst): Let hard disks sleep well and work energetically. In *ISLPED*, August 2008 Banglore, India.

71. Seung Woo Son and Mahmut Kandemir. A prefetching algorithm for multi speed disks. *Trans. High-Performance Embedded Arch. and Compilers I*, pages 317–340, 2007 Springer Berlin Heidelberg.

72. Qingbo Zhu and Yuanyuan Zhou. Power-aware storage cache management. *Trans. Comput.*, 54(5):587–602, 2005.

73. Athanasios E. Papathanasiou and Michael L. Scott. Energy efficient prefetching and caching. In *ATC*, June 2004 Boston, MA.

74. Timothy Bisson, Scott A. Brandt, and Darrell D.E. Long. NVCache: Increasing the effectiveness of disk spin-down algorithms with caching. In *MASCOTS*, April 2006 Monterrey, CA.

75. Feng Chen, Song Jiang, and Xiaodong Zhang. SmartSaver: Turning Flash drive into a disk energy saver for mobile computers. In *ISLPED*, October 2006 Tegernsee Germany.

76. Tao Li and Lizy Kurian John. Run-time modeling and estimation of operating system power consumption. *SIGMETRICS Perform. Eval. Rev.*, 31(1):160–171, 2003.

77. Seung Woo Son, Guangyu Chen, Ozcan Ozturk, Mahmut Kandemir, and Alok Choudhary. Compiler-directed energy optimization for parallel disk based systems. *Trans. Parallel Distrib. Syst.*, 18(9):1241–1257, 2007.

78. Youngjae Kim, S. Gurumurthi, and A. Sivasubramaniam. Understanding the performance-temperature interactions in disk I/O of server workloads. In *HPCA*, February 2006 Austin, TX.

79. Medha Bhadkamkar, Jorge Guerra, Luis Useche, Sam Burnett, Jason Liptak, Raju Rangaswami, and Vagelis Hristidis. BORG: Block-reORGanization for self-optimizing storage systems. In *FAST*, February 2009 San Francisco, CA.

80. Jeffrey P. Rybczynski, Darrell D. E. Long, and Ahmed Amer. Expecting the unexpected: Adaptation for predictive energy conservation. In *StorageSS*, 2005.

81. Fred Douglis, P. Krishnan, and Brian Marsh. Thwarting the power-hungry disk. In *WTEC*, 1994.

82. Igor Crk and Chris Gniady. Context-aware mechanisms for reducing interactive delays of energy management in disks. In *ATC*, June 2008 Boston, MA.

83. Ricardo Gonzalez and Mark Horowitz. Energy dissipation in general purpose microprocessors. *IEEE J. Solid-State Circuits*, 31(9):1277–1284, 1996.

84. Irving S. Reed and Gus Solomon. Polynomial codes over certain finite fields. *J. Soc. Indust. Appl. Math.*, 8(2):300–304, 1960.

85. Ilias Iliadis and Xiao-Yu Hu. Reliability assurance of RAID storage systems for a wide range of latent sector errors. In *NAS' 08*, June 12–14, 2008 Chongging, China.

86. Bianca Schroeder, Sotirios Damouras, and Phillipa Gill. Understanding latent sector errors and how to protect against them. In *FAST*, February 2010 San Jose, CA.

87. Alina Oprea and Ari Juels. A clean-slate look at disk scrubbing. In *FAST*, February 2010 San Jose, CA.

88. TPC. Transaction Processing Council. home page. http://www.tpc.org.

89. *TPC-H benchmark specification*. Technical report, http://www.tpc.org/tpch, December 2007.

90. W. Kohler, A. Shar, and F. Raab. *Overview of TPC benchmark C: The order-entry benchmark*. Technical report, 1991.

91. Priya Sehgal, Vasily Tarasov, and Erez Zadok. Evaluating performance and energy in file system server workloads. In *FAST*, February 2010 San Jose, CA.

92. Akshat Verma, Ricardo Koller, Luis Useche, and Raju Rangaswami. SRCMap: Energy proportional storage using dynamic consolidation. In *FAST*, 2010.

93. Kazuo Goda and Masaru Kitsuregawa. Power-aware remote replication for enterprise-level disaster recovery systems. In *ATC*, pages 255–260, June 2008 Boston, MA.

94. Nitin Agrawal, Vijayan Prabhakaran, Ted Wobber, John D. Davis, Mark Manasse, and Rina Panigrahy. Design tradeoffs for SSD performance. In *ATC*, pages 57–70, June 2008 Boston, MA.

95. Adam Leventhal. Flash storage memory. *Commun. ACM*, 51(7):47–51, June 2008 Annapolis, MD.

96. Feng Chen, David A. Koufaty, and Xiaodong Zhang. Understanding intrinsic characteristics and system implications of flash memory based solid state drives. In *SIGMETRICS*, pages 181–192, June 2009 Seattle, WA.

97. S. Raoux, G. W. Burr, M. J. Breitwisch, C. T. Rettner, Y.-C. Chen, R. M. Shelby, M. Salinga, D. Krebs, S.-H. Chen, H.-L. Lung, and C. H. Lam. Phase-change random access memory: A scalable technology. *IBM J. Res. Dev.*, 52(4):465–479, 2008.

98. David G. Andersen, Jason Franklin, Michael Kaminsky, Amar Phanishayee, Lawrence Tan, and Vijay Vasudeva. FAWN: A fast array of wimpy nodes. In *SOSP*, pages 1–14, 2009.

99. Alain J. Martin, Mika Nyström, and Paul I. Pénzes. ET2: A metric for time and energy efficiency of computation. *Power Aware Comput.*, pages 293–315, 2002 Springer, US.

100. Yiannakis Sazeides, Rakesh Kumar, Dean M. Tullsen, and Theofanis Constantinou. The danger of interval-based power efficiency metrics: When worst is best. *Comp. Arch. Lett.*, 4, 2005, doi: 10.1109/L=CA.2005.2.

101. Ali R. Butt, Chris Gniady, and Y. Charlie Hu. The performance impact of kernel prefetching on buffer cache replacement algorithms. *Trans. Computers*, 56(7):889–908, 2007.

102. James S. Plank. The RAID-6 liberation codes. In *FAST*, February 2008 San Jose, CA.

103. Lyndon Evans and Philip Bryant. LHC machine. *JINST*, 3(S08001), 2008.

CHAPTER 6

Cross-Layer Power Management

Zhikui Wang and Parthasarathy Ranganathan

CONTENTS

6.1 INTRODUCTION

Power delivery, electricity consumption, and heat management are becoming critical in data center environments that happen at multiple layers: in individual servers, across a collection of servers such as a blade enclosure or a server rack, at the entire data center level, or even across multiple geographically distributed data centers. In addition, at each of these layers, power control can be done at the hardware, the firmware, the operating system (OS), or the application level. A key challenge arises when these multiple controllers interfere with one another due to lack of adequate coordination. Such interference can compromise both the management objectives and the system performance. This chapter addresses the cross-layer active power management problem. We discuss the challenges and summarize the key principles to consider for coordinated power control. We then present a few examples that illustrate the problem formulation, the models, the architectures, the algorithms and policies, and the benefits of the cross-layer power management solutions.

6.1.1 Multiple Levels of Power Management

Power management is a multifaceted problem. As a result, a multitude of solutions has been proposed to address the power management problem in data centers. In the following, we examine how we can create taxonomy of these solutions across different dimensions.

First, many physical actuators are available for dynamic power management. Given that the processors usually contribute to most of the varying power consumption of servers, *dynamic processor frequency throttling* and *dynamic voltage and frequency scaling* (DVFS) are among the most popular actuators for tuning the power consumption of servers. The power consumption of the other server components, such as memory, storage, and networking, may be tuned through active tuning at run time. In some cases, when dynamic control of the status of a single component may not significantly vary the total server power consumption, the server may be set into different system power states, so the power consumption of the multiple components may be changed. An example of support for such system states can be found in the Advanced Configuration and Power Interface (ACPI) standard [1], which defines multiple active/idle states.

Ideally, a system would show good "proportionality" [2] between the active and idle states with power consumption tracking the resource usage. Servers today are still not power proportional with a significant amount of idle power consumption. Consequently, turning the entire server off is often considered one actuator when considering optimizations to save power.

Irrespective of the specific methods used for power management, a common characteristic across all these actuators is that the capacity of the servers varies with the dynamic tuning of the actuators. Since the servers are usually shared by multiple applications, tuning the states of the components may affect the capacity available for all the applications. In that context, power can be deemed as a type of "resource" to be dynamically allocated to the applications, although the relation between the actual power consumption and the capacity available to the applications could be complicated.

Second, the power of the servers can be tuned through active workload management. Traditionally, the workloads running on the servers may be controlled through mechanisms such as admission control and load balancing. More recently, workloads can be moved across different servers through virtual machine (VM) migration. Different from the dynamic provisioning of power as a resource supply, active workload management varies the power resource demand of the servers that can meet the performance requirement of the workloads.

Third, the power management problems may be formulated with different objectives, for instance, tracking, capping, or optimization. Solutions focused on average power target reduction of the electricity consumed by minimizing the power needed to achieve application performance targets. These are typically *tracking problems* in which the consumed power needs to track the resource demands of the applications. Solutions concerning peak power, on the other hand, address the provisioning of power in data centers. These are *capping problems,* which ensure that the system does not violate a given power budget. Controlling power in most systems involves changing the capacity of the servers and the application performance as well. This leads to the potential performance loss with power management. When the utility function is defined as a function of the capacity of the servers, the power consumption by the servers, and the performance of the applications, the power management problem becomes an *optimization problem* that addresses the trade-offs between the costs and benefits of both the capacity provisioning and the performance guarantees.

Finally, the implementation of the power management solutions can happen at different levels of abstraction: the chip level, the server level, the cluster level, the data center level, and even across data centers. Power management can be implemented in hardware or as software or firmware. The level at which a solution is implemented depends on several factors, including the availability of the actuators, the availability of sensors corresponding to the metrics being optimized, and the operational granularity of the solution. Typically, the software solutions have more high-level application information available, whereas the hardware solutions have more access to low-level hardware information. Other differences among power management solutions pertain to the nature of the metrics being optimized. For example, depending on the level at which the solution operates, power management can optimize a local metric or a global metric. Similarly, power management can be implemented either as a local algorithm or as distributed algorithms for global optimization.

The previous discussion points to four key high-level dimensions to reason about the diversity of power management solutions: (1) actuators to vary the power performance point of the system, (2) actuators to control workload management (and correspondingly the power demand), (3) objectives of the power management solutions, and (4) the scope of the solutions.

While the solutions individually address aspects of the power management problems in isolation, deploying them together has the potential for synergistic interactions and can better address the dynamic and diverse nature of workloads and systems in future enterprises. The need for federation, the full or partial overlap in the objective functions and the use of the same or interrelated knobs for power control across the different solutions, often at different time granularities, makes this a hard problem. In the absence of such coordination, the individual solutions are likely to interfere with one another in unpredictable, and potentially dangerous, ways.

As one example of potential conflict, the same power control knob may be utilized for multiple purposes. For instance, power states (or P states), as defined in the ACPI standard, can be utilized for both capping and tracking. If uncoordinated, the controllers for either objective can potentially overwrite the other's outputs on the P states, leading to bad application performance, power budget violations, or eventually thermal failover.

As a second illustrative example, if workload consolidation and power-capping controllers are uncoordinated, more workload than allowed by the power budget may be moved to the servers. If the power-capping controller throttles the performance to enforce the power budget, the throttled performance in turn can lead to reduced utilization, which can trigger the workload consolidation controller to pack even more workloads onto the server, which leads to a vicious cycle and system instability.

6.1.2 Cooling Management

Traditionally, the power consumption of servers, or more generally the computing systems for information technology (IT), including the servers, storage systems, and networking equipment, have been extensively studied since such power consumption is closely related to application performance. However, the IT power is only one component of the total power consumed by a data center. The other significant component is the *facility power* consumed by the cooling equipment, such as cooling fans, computer room air conditioners (CRACs), chillers, cooling towers, and the power delivery facility such as UPSs (uninterruptible power supplies) and power delivery units (PDUs). Studies showed that for every 1 W of power used to operate servers, traditionally an additional 0.5–1 W of power is consumed by the cooling and power delivery equipment that extracts the heat from or supplies power to the data center [3, 4]. This effect is characterized by a metric proposed by the GreenGrid [5], called the PUE, the power usage effectiveness. PUE is defined as the total power entering a data center divided by the power used to run the computer infrastructure within it. In the previous example, the PUE would be 1.5 to 2. The yearly electricity costs for cooling large data centers can thus reach millions of dollars.

There have been several studies to optimize for cooling that reduced the cooling power consumption while maintaining the thermal safety of the systems in data centers (typically through temperature thresholds for servers) [6, 7]. In this context, the main trade-off is the cooling capacity provisioned by the facility and the thermal performance of the servers. This trade-off is quite complicated and further adds to the complexity with multiple power controllers discussed in the previous section. Furthermore, in a data center with a raised-floor open environment, the server temperatures can be modulated in the traditional mechanical cooling system by (1) the fans inside the servers, (2) the perforated tiles on the floor that are located in front of server racks, (3) blower speeds and supply/return air temperatures of the CRAC units that provide cooling air to servers, (4) chilled water pumps and controllers within the chillers, and (5) cooling towers that provide chilled water to the entire data center. Thus, the cooling management problem itself is another federated control problem since the actuators may be located at the levels of individual servers, racks, zones of servers, or entire data centers, with inputs from, and control of, diverse sources, such as server temperature; airflow rate, pressure, and temperature; and water pressure, temperature, and flow rate. Control architectures have to be carefully designed to optimize the overall cooling efficiency of data centers while meeting the thermal requirements of the servers and other equipment.

Power and cooling management have traditionally been separated on servers and data centers, although they are closely related to each other. The power consumption of the servers dynamically varies with the workload and fluctuates considerably over both short-term and long-term time scales. Moreover, active power management varies the power consumption and, correspondingly, the cooling demand of IT systems. Without coordination with cooling control, active IT power control can possibly create new "hot spots" and reduce the overall cooling efficiency of a data center. Recent work has started exploring the increased benefits possible from cooling-aware workload management [8–13].

6.1.3 Power Capping: Another Element of Cross-Layer Power Management

Another dimension to consider in the problem of cross-layer power management is power capping [14, 15]. The power consumption of a server, a group of servers, or the entire data center may need to be capped due to

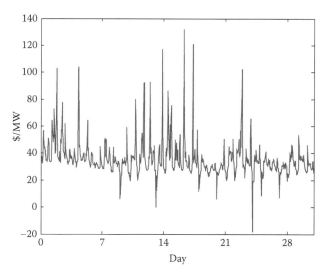

FIGURE 6.1 Hourly locational marginal pricing in northwestern Pennsylvania in March 2010. (Source: http://www.jpm.com.)

a multitude of reasons. In the case of electrical power capping, the power budget can correspond to the "capacity of the fuse" in the power supplies during operation or the power capacity available from the power grid for the entire data center. In the case of thermal power capping, the power budget can be enforced due to the limited heat extraction capacity of the server cooling fans or the limited cooling capacity of the air conditioners. Recently, there has been an increasing amount of interest in the smart grid and its impact on IT in general and data centers in particular [16]. The benefits of using a smart grid infrastructure includes a reduction in CO_2-emitting power plant construction, a more efficient and reliable electricity grid, and better customer pricing for electricity. However, the main actuators used by the smart grid include mechanisms such as time-of-use (TOU) pricing, critical peak pricing (CPP), real-time pricing (RTP), and peak time rebates. As an example, Figure 6.1 shows the hourly locational marginal price in a smart grid in northwestern Pennsylvania, where the energy price was decided through bidding that represented the availability of the power and the demand for the power. The price can vary significantly over time. To take advantage of the possible savings, power capping is required for any power management system. This leads to the introduction of power budgets at the data center level.

This chapter focuses on the control and management architecture that addresses the potential conflict of multiple interacting controllers. As is evident from the previous discussion, the combinatorial space of the multiple dimensions of power management solutions can be quite large. This chapter is not targeted at providing a *single* solution that addresses this entire space. Instead, we focus on the principles and approaches for federating cross-layer power management, followed by some examples of the architectures and the control algorithms as well as a discussion of the benefits.

6.2   PRINCIPLES OF COORDINATED FEDERATED MANAGEMENT

Several open questions exist for the design of a coordinated solution. How should the overall architecture be designed for individual controllers to interact with each other to ensure (1) *correctness* (no excessive power budget violations), (2) *stability* (no large oscillations), and (3) *efficiency* (optimal trade-off between power and performance)? How do we combine the individual tracking, capping, and optimization solutions? How do we address the lack of visibility into other controllers and minimize the need to exchange global information? Furthermore, given such a coordinated scenario, there are several implications for the design of the solution. Are all solutions equally important? Do the policies and mechanisms at the individual levels need to be revisited in the context of their interactions with other controllers? How sensitive are the answers to the nature of the applications and systems considered?

6.2.1   High-Level Problem Formulation: Constrained Optimization

Power management can be formulated as a traditional optimization problem that coordinates the supply of computing resource, power, and cooling capacity and the demand from the workload and servers themselves. Multiple objectives need to be considered for energy management in data centers, including, but not limited to, energy minimization, application performance guarantees, temperature tracking for the safety of electronic components, power capping, and the overall minimization of the total cost of ownership (TCO). The management problem can be formulated as an optimal control problem with constraints. The object is to minimize the total energy consumption of both the IT components, such as servers, networking, and storage, denoted as P_{Server}, and the facility components, such as CRACs, pumps, and chillers, denoted by $P_{Facility}$:

$$\min \int_0^t P_{\text{Servers}}(\tau) + P_{\text{Facility}}(\tau)d\tau \qquad (6.1)$$

Most of the other objectives can be represented as constraints. First, the capacity of the servers needs to satisfy the resource demand of the applications running on the servers to preserve application performance. It can be assumed that a resource utilization threshold is defined for each server, and utilization under this threshold would guarantee application performance. This means

$$Util_{Servers}(t) \leq Util_{Thresholds}(t) \qquad (6.2)$$

Second, for the safety of electronic components such as processors, memory DIMMs (dual in-line memory modules), and disks, the thermal conditions of the servers, usually represented by the server temperatures, need to be maintained below the specified thresholds:

$$T_{Servers}(t) \leq T_{Thresholds} \qquad (6.3)$$

Third, as motivated by requirements from the smart grid or any of the other power-capping requirements detailed previously, the total power consumption needs to be maintained below a predetermined or specified on-the-fly budget:

$$P_{Servers}(t) + P_{Facility}(t) \leq P_{Budget}(t) \qquad (6.4)$$

The problem defined by Equations 6.1–6.4 is a high-level generalized description of the power management solution. However, there are still a few things to consider about this general definition.

- The metrics, including the utilization and the temperatures of the servers, are functions of time due to the time-varying workload demand and dynamic thermal environments in data centers.

- Both the power consumption and the cost of energy are time sensitive, for which the integration function is needed in the cost function, Equation 6.1. Note that a more comprehensive cost function can be defined (i.e., one that incorporates the time-varying energy cost).

- The application performance requirement can be defined as a utility function, which can then also be integrated into the cost function.

- The power budget can be time varying as well, as in the case when the power needs to be clipped due to limited capacity available from the grid.

6.2.2 Understanding the System: Interaction between the Knobs and Metrics

Multiple variables and constraints are introduced in the power management problem. Given the many power management knobs that are available, the relationship between the power management actions and the variables of interest can be quite complicated. In this section, a few such examples are provided.

6.2.2.1 Power Consumption as a Function of Workload and Server Power Management

Linear models have been widely used to characterize the power consumption of servers, as in the model in Equation 6.5. This model captures the relation between the power consumption and the utilization as the result of workload management. Given that the processors are dominant in both the power consumption and the power variance of the servers, the CPU (central processing unit) utilization is taken as the representative server utilization.

$$P_{Server}(t) = a_f(t)Util_{Server}(t) + b_f(t) \tag{6.5}$$

It is worthwhile to note that the model of Equation 6.5 is already a simplification. In practice, this model can be nonlinear due to factors such as power leakage and can include dependence variables to show the effect of specific server configurations and workload types. Moreover, the power consumption is only linear in the case that the slope a_f and the intercept b_f are constant. The two parameters (a_f and b_f) can be affected by server power tuning activities such as dynamic voltage or frequency scaling and server on/off. When considering all the possible actuations that can be applied to the server, including workload management, dynamic frequency scaling, dynamic voltage/frequency scaling, and turning servers on/off, the behavior of the server can be much more complicated.

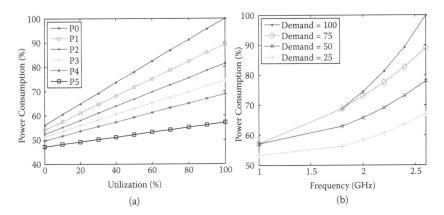

FIGURE 6.2 Power models for Server A with six P states: (a) linear function representing the relationship between the power consumption and the CPU utilization; (b) nonlinear function representing the relationship between the power consumption and the frequency.

To illustrate this further, Figure 6.2 shows the power models of an example server, called Server A, with AMD processors that have six P states (P0, P1, ... , P5) operating at core voltages/frequencies of 1.35/2.6, 1.30/2.4, 1.25/2.2, 1.2/2.0, 1.15/1.8, 1.10/1.0 V/GHz, respectively. The linear (curve-fitting) models between the power consumption and the utilization are shown in Figure 6.2a for different P states, where the utilization is defined as the ratio between the CPU capacity consumed by the workload and the maximum capacity available at a certain P state. Figure 6.2b shows the server power consumption as a function of processor frequency, converted from the data shown in Figure 6.2a. (The conversion was done based on the assumption that the highest throughput of the application supported by the server is proportional to the capacity of the cores, represented by their operational frequencies.) For a given workload demand (represented by the CPU capacity required to serve the workload when the CPU runs at the highest frequency), the server power consumption is a nonlinear convex function of the frequency. This nonlinearity is in part because when the core frequency is increased due to changes of the P state, the voltage of the processor is also increased, while the power is a quadratic function of the voltage.

As another example, Figure 6.3 shows a different server, called Server B, that has Intel processors supporting two P states, P0 and P1, operating

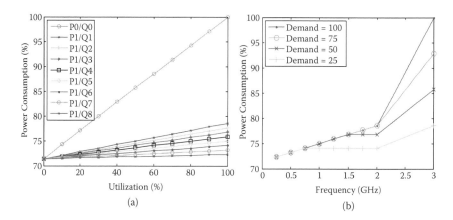

FIGURE 6.3 Power models for Server B with two P states and eight more frequency-throttling states: (a) the linear models representing the relationship between the power consumption and the CPU utilization; (b) nonlinear functions representing the relationship between the power consumption and the frequency.

at core frequencies of 3.0 and 2.0 GHz, respectively. Different from the previous server, a set of so-called Q states is defined for power management purposes. Q0 and Q1 are the same as P0 and P1. Seven more Q states (Q2, Q3, ... , Q8) are defined, corresponding to the scenarios where the processor is slowed by stopping the clock for a fraction of the time varying between 87.5% and 12.5% at step size of 12.5%. The modulation through Q states (other than Q0 and Q1) has a similar effect to frequency throttling on the power consumption and capacity available. For instance, Q3 of Server B is equivalent to a core "frequency" status of 1.75 GHz (= 2.0*87.5%) in terms of the processing capacity. By doing so, finer granularity can be provided for power tuning, and the tunable range can be extended from that achieved using only P states. Different from Server A, the power curves as functions of utilization are clustered together for the frequency-scaling states Q2 to Q8 because the actual operational frequencies are close to each other, while the voltage is not varied when the actual frequency is throttled. Figure 6.3b shows the more complicated power-frequency relationship of Server B compared to that of Server A. Take the example of a workload demand of 50% corresponding to a CPU demand of the workload that is 50% utilized when the CPU runs at P0. When the frequency is below 1.5 GHz, the CPU is always fully utilized, and the power consumption is

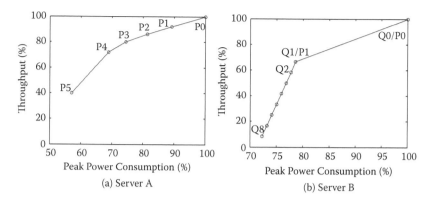

FIGURE 6.4 Throughput/power performance at different power status.

actually a linear function of the equivalent frequency. The power is even a constant when the frequency varies between 1.5 and 2 GHz. This nonlinear relationship between the power and the "virtual frequency" exists with the server whenever the server power state is varied between the Q states Q2 to Q8 (and the workload is constant). Careful consideration has to be taken when the two mechanisms, *frequency scaling (or frequency throttling)* and *voltage and frequency scaling*, are utilized for active power management since they result in very different server performance characteristics.

As argued previously, power should be deemed as one resource to be allocated. One interesting question then is how the computing performance of the server is varied along with the power. As one example, consider the throughput of the application, which in most cases is approximately proportional to the computing capacity of the processors. Figure 6.4 shows the relationship for the two servers between the highest possible application throughput and the peak power consumption when the frequency is tuned. These relationships are both nonlinear and concave. We define the metric for power efficiency $\eta = \frac{dthroughput}{dwatt}$, which means the capacity loss per unit of power (as a resource) reduction, or the capacity gain per unit of power resource increment. The nonlinearity shown in Figure 6.4a implies that the power efficiency of the servers varies along with its operational state. When the total power consumption of the servers has to be reduced, more power resource can be taken away from the servers running at the higher frequency (or the lower efficiency) so that less of the computing capacity is compromised. On the other hand, when a power resource is available for the servers, more power can be allocated to the servers running at the lower frequency that holds higher efficiency. Figure 6.4b shows the bimodal behavior of Server B again. The power efficiency of the server when operating at the states Q2 to Q8 is constant and significantly different from that of Q0 and Q1.

6.2.2.2 Application Performance Model

Although some application performance metrics (e.g., throughput) can be proportional to the computing capacity of the servers, other metrics (i.e., the end-to-end response time) can be more complicated. This section shows one example that has been discussed previously [17, 18].

Modern Internet and e-business applications are usually structured in multiple logical tiers. Each tier provides certain functionality to the preceding tier and uses the functionality provided by its successor to carry out its part of the overall request processing. Consider a general M-tier application. Assume that each tier m runs on a separate VM V_m ($m = 1, \ldots, M$). Further, assume that the CPU is the single bottleneck resource and that the CPU is the only resource to be dynamically allocated among the VMs. The end-to-end response time of a multitier application can then be calculated by aggregating the resident times over all resources (e.g., CPU, disk, network) across all tiers (e.g., web, application, and database tier). According to M/G/1/PS queue theory, the total resident time, or the mean response time (MRT), of all the requests served in all the tiers can be approximated as follows:

$$MRT = \frac{1}{\lambda} \sum_{m=1}^{N} \frac{\sum_{n=1}^{N} \beta_{nm} \lambda_n}{E_m - \sum_{n=1}^{N} \beta_{nm} \lambda_n} + \sum_{n}^{N} \alpha_n \lambda_n \qquad (6.6)$$

In the model of Equation 6.6, the MRT is broken down into two parts: the resident time on CPU resources represented by the first term and the resident time on non-CPU resources represented in the second term. The workload is assumed to have N transaction types, such as login, browsing, and bidding. Unlike workload models that assume a stationary transaction mix (and correspondingly use an aggregate request rate to characterize the workload), the model of Equation 6.6 defines the intensity of the workload as a vector $(\lambda_1, \ldots, \lambda_N)$, where λ_n is the average request rate of transaction type n during one time period. The parameter β_{nm} represents the average CPU demand of transaction type n at tier m, α_n defines the service time of non-CPU resources of transaction type n, and E_m means the CPU entitlement that is allocated to the virtual server at tier m. The model in Equation 6.6 can be extended to model the performance of the application when power is under active control. For instance, dynamic frequency scaling varies the processor capacity. However, that may not change the CPU demand β_{nm} of the application requests.

6.2.2.3 Server Temperature Model

Server temperatures, or more precisely the component temperatures of the servers, are the main metrics for server thermal performance. For a server cooled by fans, the dynamic thermal model for CPU temperature T_{CPU}, the dominant temperature sensor in most servers, can be derived by leveraging heat transfer theory for thermal resistance and energy balance via a lumped capacitance method [19],

$$C_1 \frac{d T_{CPU}}{dt} = \frac{C_2}{R(t)} \left(T_{Ambient} - T_{CPU}(t) \right) + Q(t) \tag{6.7}$$

where $T_{Ambient}$ is the ambient temperature. The variable Q represents the heat transferred per unit of time between the CPU and the ambient air. The heat can be approximated using CPU power consumption, which is modeled similar to server power as

$$P_{CPU}(t) = c_f(t) Util_{CPU}(t) + d_f(t) \tag{6.8}$$

The variable R in Equation 6.7 is the thermal resistance between the CPU temperature sensor and the ambient temperature sensor, which can be approximated as

$$R(t) = \frac{C_3}{V_j^{n_R}} + C_4 \tag{6.9}$$

where v is the volumetric airflow rate through the heat sink. The parameter n_R is determined by the shape of the thermal resistance curve as a function of the airflow rate. It is primarily related to heat-sink design and the level of turbulence in the flow, which is a function of air velocity through the heat sink. The parameters C_k ($k = 1, 2, 3, 4$) are constants related to the fluid and material properties of the air, the CPU package, and the heat sink.

Note that the models of Equations 6.6–6.9 capture the interaction among the thermal condition of servers, the workloads, the local server cooling, and the data-center-level cooling, represented by the CPU temperature, the CPU utilization, the thermal resistance, and the ambient cooling air temperature, respectively. From the dynamic control point of view, the CPU temperature is a first-order linear function of both the heat transferred and the ambient temperature. This relation has been widely utilized previously to model electronic component temperatures. However, when the fan speed is actively tuned, the time constant of the first-order system can vary significantly along with the fan speed. This complicated nonlinear and time-varying behavior of the thermal system makes it challenging to design efficient and robust dynamic controllers. It is worthwhile to note that the fan power, as a function of the fan speed, is dependent on the heat transferred, or the power consumed by the processors, and the CPU temperatures. In other words, when the total power of the fans and the servers has to be capped, it is not trivial to allocate the power budget between the fans and the servers to maximize energy efficiency.

6.2.3   Understanding the Systems: Temporal and Spatial Variance

As a few examples, the models in Equations 6.1–6.9 demonstrate the comprehensive relationship among power management actions (through resource provisioning and demand shaping) and the application or server performance, such as the temperatures and the end-to-end response times. The discussion in this section summarizes the general temporal and spatial properties of the systems that need to be considered for development of control and optimization architectures and algorithms.

- Temporal behavior of applications
  Workloads for Internet and enterprise applications are bursty. They show time-varying intensities at both the aggregate level and at the individual request level and typically exhibit patterns corresponding to different time granularities from hours, days, weeks, to months.

- Time-varying energy price
  As discussed, the price of energy can change significantly over time, in different time scales of minutes, hours, days, or seasons, depending on the source of power and the demand on power.

- Time granularities of actuators
  Due to physical limitations, and possible capacity and performance overheads, the power-tuning actuators can work at very different time

scales as well. DVFS can be enforced frequently, sometimes even at the granularity of a few nanoseconds. Their performance overhead is low, and these solutions can be implemented at the hardware, firmware, or software level. Admission control and load balancing can be executed at very fine granularities (e.g., per request), while VM is typically done at longer time granularities due to its effect on the delay, computing capacity, and networking bandwidth consumption. Turning on/off servers can take much longer, and this actuator may not be invoked frequently due to the capacity loss of the servers and the effect on the application performance.

- Time properties of metrics
  The SLOs are defined against certain time scales. For instance, the MRT values are sensitive to the time periods for averaging, and an MRT threshold is always defined against the time period. As another example, power budget violations may be monitored at the granularities of microseconds, seconds, minutes, or even hours, depending on the objectives of the power capping. Similar leeway exists with the temperature threshold, defined in the specification of the components.

- Time constants of the systems
  The time constants represent the responsiveness of the systems to changes/disturbances applied to the systems. The power consumption of the server components changes almost immediately following workload changes or tuning off the power states. The end-to-end response time can have a much larger time constant due to the queuing processes along the travel path of the request. CPU temperatures can take a few seconds to track the changes in the workload but take longer to respond to the changes in fan speeds due to the relatively slower thermal processes in the heat sink. At the data center level, the intake air temperatures of the racks take tens of minutes to converge on changes in CRAC operation conditions.

- Scope variability across actuators and optimizations
  Server power tuning affects all the VMs and the applications hosted on the server, while workload management changes the power consumption of the servers hosting the workload and the performance of the applications sharing the servers with the workload. Temperature thresholds can be defined per component, per server, per rack of servers, per thermal zones, or per data center (e.g., the threshold of the returned air

temperatures of the CRACs). The performance metrics of the applications are typically defined per application. A few performance metrics may be introduced for multiple applications (e.g., for performance differentiation purposes). Electrical power capping can be enforced per component, per server, per rack, per multiple racks that are served by one PDU, or the whole data center. Finally, energy minimization can be done against the constraints of a single application up to the workloads across multiple geographically distributed data centers. It is worthwhile to emphasize that, due to the interaction among the actuators, metrics, and the optimizations, most actuators can be utilized for multiple objectives. For instance, P-state tuning can be used for both power efficiency and power capping, and workload management can help with power efficiency, power capping, and temperature control.

- Heterogeneity
  In addition to temporal variance and variability in scope of influence, heterogeneity of workloads and systems needs to be considered. Especially, different applications have very different behavior, and the performance requirements can vary by workload type. Similarly, servers are heterogeneous in terms of capacity, power performance, and energy efficiency. For cooling management, the cooling efficiency in the data center is location dependent due to physical design issues such as layout of the racks as well as workload distribution.

6.2.4   High-Level Principles of Architecture Design

When designing an architectural framework for cross-layer power management, it is worthwhile considering the physical and functional topologies already present in the data center. For example, the physical IT infrastructure is typically hierarchical: The data center is composed of rows (aisles) of racks, each rack hosts a number of servers, and each server runs a set of VMs (in a virtualized environment). Inside the servers, there are layers, such as the hardware, the firmware, the hypervisor, the VMs, and the applications, typically visualized as a stack. The servers can be grouped together for different purposes. As one typical example, the VMs can only be migrated in a domain of the servers with shared storage. In another example, the networking structure in a data center can apply constraints to the scope of the actuation, such as VM migration, as well.

Other types of architecture also exist in the data center. Each application can span multiple physical servers. For instance, a typical multitier application has a web tier, an application tier, and a database tier; each tier can be hosted in multiple physical or virtual servers. The cooling infrastructure of a data center is not exactly hierarchical. Inside a typical open and air-cooled data center with raised floor, the cool air can be provided by the CRAC units to zones of the data center and is then sent to the racks through local cooling actuators (e.g., perforated tiles), and the cooling air is then pulled through the individual servers by the cooling fans inside the servers. However, the cooling capacity of the data center may be provisioned by multiple resources; for example, it can come from a water economizer or outside-air cooling or some energy storage systems. The networking and storage systems have different, often more complicated, topologies.

Given the diverse topologies of physical and logical infrastructure in the data center, there is no single architectural approach that works best for power management at the data center level. However, there are still a few principles that may be generally applicable.

- Federation
  Federation is needed in the presence of the temporal and spatial variances. Multiple controllers are needed to address the individual pieces of the problems; however, they are also required to coordinate with each other.

- Formal optimization
  Optimization is necessary to address the multiple objects, the multiple constraints, and the interactions among the actuators, metrics, and subproblems. However, a single gigantic optimization process is not reasonable due to the temporal and spatial variation discussed in the previous section. Such a process is infeasible because of both the complicated behavior of the systems themselves (e.g., nonlinearity, discontinuity, the large parameter space) and the unavailability of models that are accurate enough. In addition, even the most detailed models still have the potential for inaccuracies, not all the dynamics of a real-time environment can be captured in mathematical models, and the parameters of the models can vary with time. As one set of critical inputs to the models, the capacity, power, and cooling demands for the servers are usually not predictable due to bursty workload behaviors. The computing capacity overhead of the optimization process can sometimes be significant as well. On the other hand, optimization may be applied for some subproblems (i.e., VM migration and capacity

planning when reasonably accurate models and metrics are available), and the inaccuracy of the decisions may be compensated by other lower-layer controllers.

- Feedback control
Feedback control techniques are very helpful to deal with (1) complex systems that may not be captured by mathematical models, (2) disturbance to systems that are not under control, and (3) unpredictable workload variance. However, to apply a control-theoretic approach to power management, the problem will often need to be formulated as a tracking problem, with one or more metrics to be driven to some reference values at the steady states or some reference trajectories. One application of feedback control is to manage the end-to-end response time of the application when the response time is to be maintained at a given reference level upon changes of the workload. The reference level can be the response time threshold defined in SLA or derived through higher-level cost optimization.

- Hierarchical control
Hierarchical control can be one natural choice to deal with some problems. For instance, to meet the power budget of the data center, the budget can be enforced by clipping the group power consumption of a cluster of servers, grouped based on PDU connections. The power budget of each group is then enforced by capping the power consumption of the servers. Hierarchical control can be implemented in cooling management as well. For instance, the CRAC units provide cool air to the zones of the data center, which is then redistributed to the racks through perforated tiles. In that case, the "zonal" cooling controller responds to the "aggregate" local cooling demand. For more complicated cases, a common guiding principle is to enable coordination, wherever possible, by connecting the actuation at one layer to the inputs at another layer. This allows the feedback controller to react to (and learn from) interactions across controllers (e.g., through the changes of its reference value) the same way as it would react to disturbances.

- Minimal interfaces
An important principle in cross-layer power management is to minimize the number of explicit changes in the individual controllers for coordination. This avoids the performance issues around global

information exchange or the availability issues around a centralized arbitration model.

- Formal rigor, flexibility, and extensibility
  Stability and performance have to be guaranteed for the feedback controllers, especially in the face of changing workload demand and interacting controllers. In addition, a cross-layer power management solution needs to be flexible, allow different deployment scenarios, and work well with the dynamic nature of enterprise data centers. Changes of workload behavior, system models, controller policies, time constants, and the like are all to be accommodated. Finally, the architecture should easily be extended to other classes of controllers and other specific implementations.

6.3 CROSS-LAYER POWER MANAGEMENT ARCHITECTURES

In this section, we discuss how the broad principles discussed can be applied in the context of specific cross-layer power management solutions. We present four examples: (1) coordinated power capping and power efficiency controllers (ECs) in a set of servers; (2) a cross-layer power management solution that includes server-level and data-center-level power management, including VM power management with capping and ECs; (3) unified power and application SLA management; and (4) unified power and cooling management.

6.3.1 Coordinated Server-Level Power Management: Power Efficiency and Power Capping

As the first case, Figure 6.5 represents the control system architecture for a group of servers for the purposes of server-level power efficiency and power-capping management. It introduces three controllers to the servers: an EC that adapts the power consumption of the individual server to track the demand of the workload, a server capper (SC) that throttles the power consumption of the individual server, and a group capper (GPC) that throttles the total power consumption of the group of servers. To improve the power efficiency of the server, it is considered as a "container" that should be used at a desired fraction of its available capacity, or "utilization," notated as the reference (refUtil). Regulating resource utilization at its reference drives the EC to dynamically "resize the container" by varying the processor voltage/frequency through P-states tuning. This allows the power consumed to adapt to the resource demand the workload places on the server

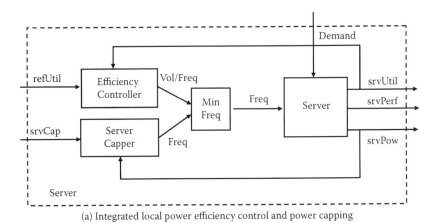

(a) Integrated local power efficiency control and power capping

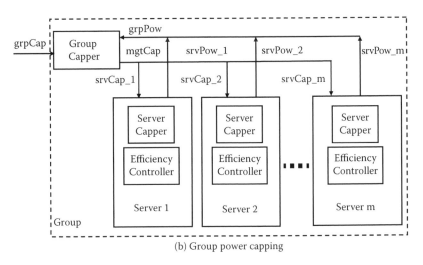

(b) Group power capping

FIGURE 6.5 A group of servers with active power efficiency control and power capping.

in real time. The target utilization is set up for performance. It is usually lower for the workloads with bursty demand than for workloads with less variance.

The group power capping is enforced through two controllers. At the group level, the GPC collects the power consumption of the individual servers (srvPow) and the total power consumption of the group (grpPow), based on which it distributes the group power budget (grpCap) to the group members (srvCap). At the server level, the SC measures the per server power consumption (srvPow), compares it to its power budget (srvCap), and drives the server power consumption to the budget level by changing the processor voltage/frequency through P-states or Q-states tuning.

The two-layer control design for power capping provides a peak power management architecture that is easy to implement, reliable, and distributed. The SC can be implemented in the local servers. The GPC can be located at different levels, for instance, the blade enclosures, the racks, or even the data center. The two controllers work in different time scales. The control intervals of the SC can be seconds. The GPC needs to communicate with the individual servers. It runs at lower frequency (e.g., in tens of seconds).

The EC runs in the individual servers. Among the three controllers, it runs at the shortest time scale (e.g., 100 ms), so that it can track bursty workload demands. Both the EC and the SC tune the operational voltage/frequency of the processors. To avoid confliction, the lower one of the two frequency outputs from the two controllers is finally actuated. This implies that, when the server power is below the server budget, the EC is dominant. But, if the server power goes above the budget, the SC will throttle the power consumed even if the utilization is above the reference, which can potentially compromise the application performance.

6.3.2   Multilayer Power Management Solution

In this section, we discuss one case with multiple layers of power controllers and more power management actuators besides that of frequency scaling/throttling. The individual controllers are representative in data centers currently that have many blade enclosures and servers. Most of the controllers are now available commercially.

Figure 6.6 shows the architecture with five controllers that are representative for their diversity: An EC optimizes per server average power consumption; a local power capper (LPC) implements thermal power capping; as in the previous section, an enclosure power capper (EPC) and a group power capper (GPC) implement (thermal) power capping at the blade enclosure and rack or data center levels, respectively; a global controller seeks to reduce the average power consumed across a collection of machines by consolidating workloads and turning unused machines off. There are power budgets in the server, enclosure, or group levels that can be provided by system designers or data center operators based on thermal budget constraints or determined by high-level power managers.

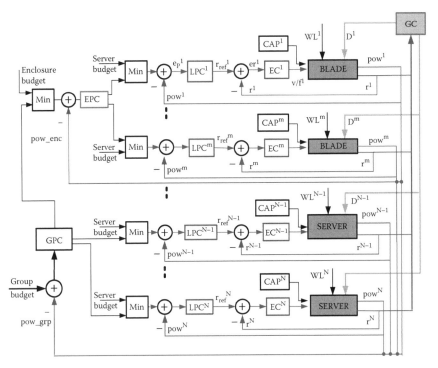

FIGURE 6.6 One architecture for integrated power management.

Figure 6.6 represents one example of a set of *cascaded* controllers for both power efficiency and power capping. The innermost level is the EC, acting similarly to the EC in Figure 6.5. It regulates the CPU resource utilization at its reference by varying the voltage/frequency of the processors. The LPC caps power at the blade/server level and is implemented as a controller nested around the EC. A key aspect of the design is that the target utilization level r_{ref} is used as the actuator for the LPC (rather than P-state tuning). In the event of a power budget violation, the controller increases the r_{ref} input to the EC, which in turn responds by going to lower P states, enabling the power budget to be met. Working in a reactive way, this approach may lead to transient budget violations, but the controller bounds the time on such violations. As discussed, this is acceptable in a thermal power capper. An optional electrical power capper (CAP) can be implemented in parallel to the EC as shown in the figure and can respond to the electric power budget violation immediately.

The EPC implements enclosure-level power capping. For each control epoch, the EPC controller monitors the total power consumption of the blade enclosure and compares it with an enclosure-level power budget. Based on the comparison, the controller assigns power budgets for the next epoch to all the individual blades in the enclosure. The LPC controller in each blade uses the minimum of the power budget recommended by the EPC and its own local power budget as its input reference. The actual division of the total enclosure power budget to individual blades is policy driven, and different policies (e.g., fair share, FIFO [first in, first out], random, priority based, history based) can be implemented. Essentially, the communication between the layers happens through the power budget settings and the measurement of the consumed power.

The group-level power capping, implemented by the GPC, works fairly similarly, but at either the rack level or the data center level and with different time constants. The actual power consumption of the group is compared to the group power budget, based on which the power budgets are assigned to all the next-level servers and blade enclosures. As before, within the EPC and the LPC, the lower of the GPC's recommendation and the local budget values is chosen.

The last element of the architecture is the global controller (GC). Different from the previous controllers that tune the power status of the servers or change the power budget references, the global controller is to distribute the workloads through ways such as load balancing and VM migration. It reshuffles the distribution of the workloads, consolidates the workloads onto fewer servers when the demand is low, and turns off idle servers so that a minimum amount of power will be consumed by the servers while still meeting the performance requirement of the applications. When the demand increases, the global controller will redistribute the workload and turn on servers if needed.

The global controller is not cascaded with the other controllers as the power-capping controllers do. It manipulates the resource demand (such that the power demand also is manipulated) on the servers based on historical data of the workload and the servers while the other controllers tune the computing and power resource supply. The GC controller works in longer time intervals so that the demands can be satisfied through the lower-layer controllers. On the other hand, the GC controller has to be aware of the constraints the lower-layer controllers will apply. For instance, when enforced by the power-capping controllers, the local power budgets of the groups or the individual servers clip the capacity available to the servers as well. These budgets have to be respected by the GC controller when the decision on workload distribution is made. Otherwise, more workloads than

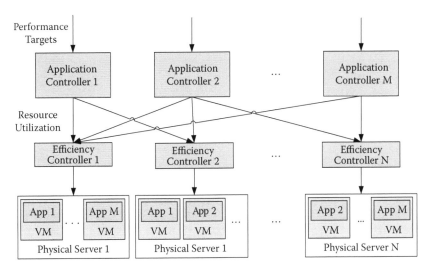

FIGURE 6.7 Performance-driven resource allocation.

allowed by the budgets can be migrated onto the servers, which will result in additional performance loss.

6.3.3 SLA Power Management

In the previous two cases, we did not explicitly consider end-to-end performance metrics such as the response time of the applications. As shown in the performance model of Equation 6.6, the response time of an application with multiple tiers can be tuned through the utilization of the VMs that host the tiers of the application, defined as the ratio between resource consumption $\sum_{n=1}^{N} \beta_{nm}\lambda_n$ and resource entitlement E_m. Figure 6.7 represents one such architecture that extends the power efficiency control to application performance management. The application controller (for each application) sets up the utilization targets of the VMs that host the individual components of the application, driven by the performance target of the application itself. The utilization targets of the individual VMs can be tuned through feedback control techniques or feed-forward control based on the performance models (e.g., the model shown in Equation 6.6) and real-time measurement of the workload parameters, such as the intensities and resource consumption of the application. The EC (for each physical server) sizes the physical server through dynamic tuning of the processor frequency to minimize the power consumption while meeting the aggregate resource demand of all the VMs running on the server. The EC in this case is also responsible for allocating the CPU shares available to the VMs to meet the individual resource demand. If the aggregate demand exceeds the maximum capacity of the server, the EC needs to do arbitration to allocate the CPU shares based on some fairness criteria or predefined policies such as priorities of the applications.

6.3.4 Unified Power and Cooling Management

The three cases discussed in the previous sections did not consider the thermodynamics of the servers and the data centers yet as described in the examples in Equations 6.7–6.9. Compared with the server power consumption or utilization metrics, the temperatures of the servers are relatively slower to respond to the changes in the workload, tuning of the power states of the servers, or changes to the cooling actuators, such as the server fans. The dynamics of the server temperatures are more complicated than the other metrics. However, the architectures discussed in the previous sections can still be extended to accommodate the thermodynamics and the thermal requirement of the servers, which is represented by the temperature thresholds. As one example, we describe how one could extend the architectures shown in Figures 6.6 and 6.7 for power management of a group of servers, each with its own cooling fans; a set of blade enclosures, each with a set of blades and cooling fans; and a set of applications, each with multiple tiers running on multiple VMs that can span multiple blades or servers. The key components include the following:

- An application controller for each application that decides the resource demand of the VMs that host the individual tiers of the applications

- An EC for each blade or server that minimizes the power consumption of the blade or server while meeting the demand of the VMs running on the blade or server

- A fan controller for each server or each blade enclosure that tunes the fan speeds dynamically to maintain the server or blade temperatures below their thresholds while minimizing the total fan power consumption

- A server LPC for each server that maintains the total power consumption of both the server and its cooling fans below a given threshold through DVFS

- A blade LPC for each blade that maintains the power consumption of the blade below a given power budget through DVFS

- An EPC for each enclosure that maintains the total power consumption of the enclosure (including the enclosure itself, the blades, and the cooling fans) below a given power budget through setting of the power budgets for the individual blades

- A GPC for the group of the servers and enclosures that keeps the total power consumption of the group below a given threshold through setting of the power budgets of the servers and the enclosures

- A GC for the group that migrates VMs to consolidate workloads and turns servers on/off when needed.

6.4 CROSS-LAYER POWER MANAGEMENT SOLUTION DESIGNS

The power management architectures define the interactions between the subsystems that are located at different layers and are under active control or optimization. We discuss the management objectives, the control actuators, the metrics to be observed, and the possible inputs/changes/disturbances to the individual subsystems. Many techniques are available to achieve the management objectives. In this section, we discuss two types of techniques often utilized to address the complicated management tasks, especially during the operation of the systems: feedback control and optimization. To ground the discussion, we focus on the first two architectures of the previous section. We end this section with a discussion of evaluation studies showing the benefits and learnings from these approaches.

6.4.1 Feedback Control and Case Studies

6.4.1.1 An Adaptive Power Efficiency Control Algorithm

In the architecture shown in Figure 6.5, the power consumption of the servers is minimized while the total power consumption of the group is capped below a threshold. There are feedback loops shown in the figures for both the power EC and the power cappers. In this section, we give a few examples that show how to apply feedback control techniques in the design of the controllers.

Figure 6.8 describes an adaptive feedback algorithm for the server EC. The controller runs in discrete times from interval to interval. Each interval starts with data sampling (e.g., measuring the CPU utilization, which is

At the beginning of the j_{th} control interval,
1) Poll the average CPU utilization in the previous interval
  $util(j-1)$;
2) Calculate the new operation frequency as following
  $f(j) = f(j-1) - \lambda \frac{f_Q(j-1)util(j-1)}{refUtil}(refUtil - util(j-1))$;
  $f(j) = \min(f_{max}, \max(f_{min}, f(j)))$ (8.1)
3) Quantize $f(j)$ to the P - state operation frequency $f_Q(j)$;
4) Enforce the new P - state if $f_Q(j)! = f_Q(j-1)$.

FIGURE 6.8   An adaptive power efficiency controller.

defined as the ratio between the CPU consumption and the allocation). The CPU allocation is constant in the previous interval; however, the CPU consumption can be bursty due to the changes of the workload. In reality, the CPU consumption (or the CPU utilization) can be sampled more frequently, and the average is taken for the controller. Equation 8.1 in Figure 6.8 implements a feedback controller for the new CPU operational frequency. It is an integral controller for which the change of the operational frequency is proportional to the utilization error (i.e., the deviation of the utilization from its reference). When the workload demand or the utilization reference is varied, the integral operation pushes the disturbed utilization to converge to its reference by tuning the operational frequency of the processors so that the exact amount of the capacity is provisioned to meet the demand of the workload.

The integral controller defined in Equation 8.1 in Figure 6.8 has a self-tuning integral gain that is proportional to the resource demand represented by the utilized capacity, that is, $f_Q(j-1)util(j-1)$, when the variable $f_Q(j-1)$ represents the operational frequency in the previous interval. The variable gain implies that the controller can respond faster to the error when the demand is relatively higher. On the other hand, the controller will reduce the frequency slowly when the workload demand is reduced and leave capacity space for a sharp increase in the workload.

The feedback control and the adaptive gain help the controller to adapt the power consumption of the servers to the varying workload demand. However, it will cause oscillation and performance loss if the gain parameters are not carefully designed. Mathematical analysis showed that the system under control is locally stable under the condition that $\lambda \in (0, 2)$.

A sufficient condition can also be derived that the system is globally stable if $\lambda \in (0, \frac{1}{r_{ref}})$. A sketch of the proof for global stability can be found in Reference 20.

"Windup" of the output can happen to the integral controller if the actuator is capped in reality but the output of the controller is not. For instance, when the demand is low and the capacity is still more than enough to meet the demand even if the frequency is pushed to the lowest one, the utilization error will always be positive, pushing the outputs of the controller lower and lower. When the demand is increased and more capacity is needed, it will take the controller a long time to pull back its output and cause significant performance loss due to the sluggish response. Step 2 of the algorithm implements the simple "antiwindup" mechanism for the integrator: The output of the controller is further capped in the range defined by the highest and lowest operational frequencies of the processors.

The processor operational frequency is assumed to be continuous in Equation 8.1 in Figure 6.8. However, the actual running frequency is discrete. In step 3, the output of the controller is quantized to the discrete frequency of the P states through rounding up or down to the closed one. Even though the actual utilization can oscillate between two values due to the quantization and feedback control, the average value should converge to the reference in a time period that is longer than the control interval.

6.4.1.2 An Adaptive Server-Capping Algorithm

The nonlinear relationship between frequency and power consumption imposes challenges for the design of the server power-capping algorithms. Figure 6.9 shows one such example on the power consumption and throughput for Server B when the server is under control of a proportional-integral-derivative (PID) power capping algorithm. The processor frequency is throttled to tune the power consumption based on the error between the power consumption and the power cap. The power budget (or cap) is varied between 70% and 100% of the maximum power. The workload demand is high enough so that the server is always fully utilized. Figure 6.9a implies that the average power consumption is kept below the budget for all the budget levels. However, when the power budget is set at 82%, as in point (2) of Figure 6.9b, the actual throughput was much lower than the ideal one. This deteriorated efficiency is caused by the nonlinearity between the frequency and the power, as illustrated in Figure 6.4. More specifically, the gain from the frequency to the power consumption (when the demand is very high) varies across the operational regions. Thus, it is challenging to

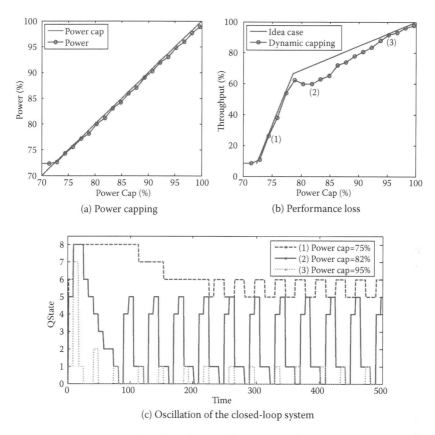

FIGURE 6.9 Power and application performance of Server B under control of a PID controller.

find one controller gain that can stabilize the closed-loop system in the entire operational range. Figure 6.9c provides more insight. For the power budgets of 95% and 75%, corresponding to points (1) and (3) in Figure 6.9b, the power state in the steady state was tuned between two adjacent power states alternatively, Q0 and Q1 for the case with the 95% budget or Q5 and Q6 for the case with the 75% budget. However, when the budget was set to 82%, the power state oscillated between Q0 and Q6, although the equilibrium should be between Q0 and Q1. In other words, the controller was too aggressive at those points. Pushing the power states to those with lower performance (i.e., Q4 and Q5) compromised the processing capability of the server, although the mean power could still be kept below the budget.

At the beginning of the k_{th} control interval,

1) Polls the average power consumption in the previous interval $srvPow(k - 1)$;

2) Runs the integral control:

$err(k) = (srvCap - srvPow(k - 1))/(MaxPow - MinPow)$

$qv(k) = qv(k - 1) + K_I * err(j)$

$if(qv(k) > 100) \ qv(k) = 100; \quad if(qv(k) < 0) \ qv(k) = 0$

3) Maps $qv(k)$ to that of the expected $Q -$ state

if(qv < qv_TH)$indQState(k) = $floor(Gain_L $*$ qv + nqStates);

else$indQState(k) = $floor(Gain_H $*$ (qv_TH $- 100$));

4) Enforce the new Q - state if $indQState(k)! = indQState(k - 1)$.

FIGURE 6.10 An adaptive power-capping controller.

Is it possible to reduce the controller gains to damp the oscillation? The answer is yes. However, the transient performance can be compromised if the gain is too low. As can be seen from the initial transient processes when the budget was set to each of the three levels, the controller responded quickly to the change of the budget to 95%; the setting time was much longer for the change of the budget to 75%. This implies that reducing the controller gains further deteriorates the transient processes when the budget is relative higher. This is the trade-off between stability and responsiveness to be addressed for any feedback controller design. In these cases, the nonlinearity of the systems makes the configuration more challenging than usual.

Figure 6.10 provides one detailed control algorithm for capping the power consumption of Server B through some linearization mechanisms. One integral controller was applied as in step 2, where the output was clipped in the range [0, 100], which was then mapped to the index of the final Q states in step 3. This output value was then enforced in step 4 if it was different from the current power state. The critical design of the algorithm is the piecewise linear transformation from the controller output qv to the operating frequency, represented by the index of the power states *indQState*. As shown in Figure 6.11a, the controller output was no longer mapped uniformly to the frequency. It was multiplied by the multiplier *Gain_L* when it was below a threshold *qv_TH*. Otherwise, it was multiplied by a lower value *Gain_H*. The relationship between the controller output and the power consumption without the transformation is represented

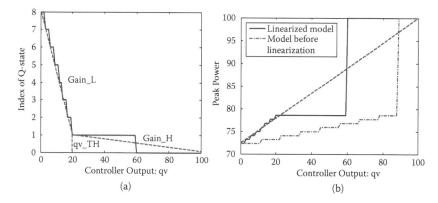

FIGURE 6.11 Nonlinear mapping and linearization for Server B: (a) nonlinear mapping from controller output to the frequency; (b) linearized relationship between controller output and power consumption.

by the dashed dot line in Figure 6.11b, demonstrating much larger gain when the output was closer to the high end. With the piecewise linear transformation, the relationship between the controller output and the peak power consumption was linearized, as shown by the dashed line. Note that the actual variables were quantized to those of the power states.

Note also that in the integral controller defined in step 2, the error was normalized by the maximum tunable power range of the server, the difference between the highest and lower power consumption of the server when it was active. This normalization helps the controller to adapt to the servers with different power specifications. It is also worthwhile to note that the integral controllers, instead of the general PID controllers, were utilized for both power efficiency control and power capping. This is mainly because the server utilization and power consumption of the server can respond to the changes of the workload and the reference values almost immediately (compared with the control intervals). The negligible time constants in both cases imply that the integral controller was able to provide good enough robustness and fast enough responsiveness to the closed-loop systems.

To evaluate the algorithm in Figure 6.10, the same experiments as those for Figure 6.9 were repeated with the results shown in Figure 6.12. Again, the power consumption was maintained below the budget. Moreover, the maximum throughput was achieved for all the budget levels. The time series in Figure 6.12c demonstrates that the adaptive algorithm eliminated the oscillation shown in Figure 6.9c for the case with the power cap of 82%.

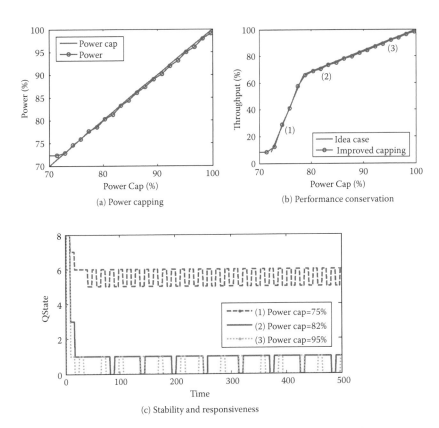

(a) Power capping

(b) Performance conservation

(c) Stability and responsiveness

FIGURE 6.12   Convergence of adaptive power-capping control.

Furthermore, the transient processes show that the controller responded to the changes quickly for all the three different budget levels.

6.4.1.3   Group Power Capper

The goal of the GC is to distribute the group power budget to the individual servers by setting up their power budgets, which are then enforced by the SCs. Many policies are available for redistribution of the budget. Figure 6.13 describes one proportional sharing policy, that is, the group power budget is allocated to the individual servers proportionally to their power consumptions.

Although simple, the algorithm in Figure 6.13 still holds reasonable properties. It is easy to find that, at equilibrium, the budgets of the servers are proportionally utilized; the budget of a server is increased when it consumes more power if the power consumptions of the other servers are unchanged.

At the beginning of each l_{th} control interval,
1) For each server m, poll the average power consumption in the previous interval $srvPow_m(l-1)$;
2) Poll the total power consumption of the group grpPow(l - 1);
3) Distribute the group power budget to the servers proportionally $\text{srvCap}_m(i) = \frac{srvPow_m(l-1)}{grpPow(l-1)} grpCap.$

FIGURE 6.13 A group capping controller.

In the special case when the budget of a server is depleted, the higher budget is allocated to the server than in the previous interval if there were extra budget available. However, it is not always guaranteed that the budgets of the individual servers are enforced by the SCs. For instance, overshoots could happen since the power capping in the servers is defined as a tracking problem and implemented as an integral controller. Implementation of the policies needs to be taken care of in these exceptional cases.

6.4.2 Optimization and Case Studies

In this section, we consider the global controller of Section 6.3.2 and discuss the problem definition and a few issues to be addressed in the implementation.

6.4.2.1 Optimization Problem for Power Management

The goal of the global controller is to consolidate the workloads (through VM migration) and turn on/off the idle servers if needed to minimize the total power consumption of the servers while maintaining the performance of the applications and the constraints on the power budgets in the group level, the server level, and the blade level. Figure 6.14 presents one formulation of the constrained optimization problem that can be solved by the global controller in each epoch.

Figure 6.14 is a standard constraint-based optimization problem as discussed elsewhere. In this formulation, it is assumed that there are a total of m servers and blades and n VMs hosting the application components. The decision variable is an m-by-n matrix X that maps n VMs to m servers or blades, the elements of which take integer values of 0 or 1 as defined in Equation 14.8 in the figure. The objective is to minimize the total power

$$\min \sum_{i=1}^{m} pow_i \tag{14.1}$$

$$s.t. \quad r_i = \sum_{j=1}^{n} X_{ij} U_j (1 + \alpha_V) \leq \bar{r}_i, \quad i = 1, 2, ..., m \tag{14.2}$$

$$pow_i \leq CAP\_LOC_i, i = 1, 2, ..., m \tag{14.3}$$

$$\sum_{i=1}^{m} M_{iq} pow_i \leq CAP\_ENC_q, \quad q = 1, 2, ..., l \tag{14.4}$$

$$\sum_{i=1}^{m} pow_i \leq CAP\_GRP \tag{14.5}$$

$$\sum_{j=1}^{n} X_{ij} = N, i = 1, 2, ..., m \tag{14.6}$$

$$\sum_{i=1}^{m} \sum_{j=1}^{n} \alpha_M \left| X_{ij} - X_{ij}^0 \right| \leq OH_M \tag{14.7}$$

$$X_{ij} = \begin{cases} 1, & \text{if VM } j \text{ is on server } i \\ 0, & \text{otherwise} \end{cases}, \tag{14.8}$$

$$i = 1, 2, ..., m, \quad j = 1, 2, ..., n$$

$$r_i = \min(1.0, \sum_{j=1}^{n} X_{ij} U_j (1 + \alpha_V))_i, \quad i = 1, 2, ..., m \tag{14.9}$$

$$pow_i = \begin{cases} c_0 r_i + d_0, & \text{if } \sum_{j=1}^{n} X_{ij} \geq 1 \\ 0, & \text{otherwise} \end{cases} \tag{14.10}$$

FIGURE 6.14 Constraianed optimization problem for the global controller.

consumption of all the servers or blades (Equation 14.1 in the figure), considering the constraints of the server utilization (Equation 14.2) as well as local-, enclosure-, and group-level power budget constraints (Equations 14.3, 14.4, and 14.5 in Figure 6.14).

A few issues have to be addressed to solve this optimization problem.

Inputs to define the current system architecture are needed. The variables m, n, and l define the number of the blades and servers, VMs, and enclosures. It is assumed that all the blades and servers belong to one group. The relation between the blades and the enclosures is represented by the matrix M_i^q. That is,

$$M_{iq} = \begin{cases} 1, \text{if blade } i \text{ belongs to enclosure } q \\ 0, \text{otherwise} \end{cases}, q = 1, 2, ..., l \tag{6.10}$$

The matrix X_{ij}^0 defines the mapping between the VMs and the blades/servers at the beginning of the global controller control interval that is needed as the inputs of the global controller.

The inputs to the controller in every interval also include the mean resource utilization values of the individual VMs, U_j, $j = 1, 2, \ldots, n$. These can be taken as the prediction for the resource demand of the workloads in the next global controller interval. In reality, the utilization of the VMs can be time varying due to both the varying workload demand and active server power management. The utilization inputs to the global controller need to be converted to values that are comparable to each other. For example, two servers with 100% utilization are not comparable if one of them is at the highest power state and the other is at a lowest power state; the latter is a potential candidate for consolidation, while the former is not. This problem can be addressed by having the global controller consider the *real* utilization, that is, the utilization when the power state is set to that with the highest frequency, instead of the *apparent* utilization. Simple models can be used to translate apparent utilization to real utilization when the power state is known, and the capacity is assumed proportional to the operational frequency of the processors. By doing so, the aggregate utilization of the servers can be estimated by adding all the utilization values of the VMs as in Equation 14.9 in Figure 6.14. Note that the utilization levels are capped by 100%, and capacity loss can happen when the server is overloaded. For heterogeneous servers, the utilization values need to be the absolute capacity demands of the workload that are comparable between servers with different configurations.

As another set of inputs to the global controller, the power consumption values of the servers can be estimated using the power models at P_0, the power state with the highest frequency (Equation 14.10 in Figure 6.14). If no VM is hosted by the server, the power consumption is zero since it is expected to be turned off. In a more practical case, some servers can always be left idle but running as a "buffer" against sharp changes of the workload demand. The size of this buffer, however, will have to be carefully chosen.

Both the resource demand measurement and power estimation are done based on the average utilization of the workloads in the past global controller control interval with the assumption that the processors work with the highest frequencies, ignoring the local power management. By doing so, we can simplify the problem definition. On the other side, this is reasonable since the global controller varies the aggregate resource demand on the servers in a relatively longer time period. This leaves the local resource management to be done by the local controllers to meet the demand of the individual workloads in much shorter time scales. In most cases, the utilization threshold $\bar{r}$ is far below 100%, leaving capacity available for the bursty workload in the shorter time intervals. In some cases, given the saturating nature of resource utilization metrics, the throttled performance can be misinterpreted by the global controller as extra space for consolidation. Feedback mechanisms need to be introduced to prevent the vicious cycle.

However, the global controller does need to be aware of the approximate budget caps at the local levels. Otherwise, a conventional design can aggressively pack workloads onto a server, which in turn can compromise the statistical load variations that the LPC, EPC, and GPC expect, leading to more aggressive performance throttling. This problem can be addressed by having the global controller (1) be aware of the approximate power budgets at the various levels and use them as constraints in its optimization and (2) be aware of power budget violations at individual levels and use them to vary the aggressiveness of consolidation. Obtaining information on the former is fairly straightforward; either machine specifications or approximate estimates can be used. For the latter, the individual capping controllers are required to expose information on their power budget violations externally. This is still reasonable and can be done by extending current Common Information Module (CIM) models exposed through Distributed Management Task Force (DMTF) interfaces. An alternate approach is to determine a proxy for the power budget violations using P states and performance violations, but this approach is likely to have more hysteresis compared to using CIM interfaces.

Overhead has to be considered when a VM is hosted on a server or migrated from one server to another. In problem formulation as in Figure 6.14, the virtualization overhead is modeled proportional to the total utilization of the VMs, represented by the weight factor α_V in Equation 14.9 in Figure 6.14. VM migration can consume CPU, memory, and network bandwidth. This overhead is represented by the factor α_M as in Equation 14.7 in Figure 6.14. Note that it is not intuitive to model the migration overhead since it may be related to application-level performance, such as delay, and the migration process has only transient effect on the application. In the case when $\alpha_M = 1$, the constraint (Equation 14.7 in Figure 6.14) sets a limit on the number of migrations allowed in every global controller control interval. Equation 14.6 in Figure 6.14 enforces the constraint on the maximum number of VMs that can run on each server. Other similar constraints can be assumed for other limited resources, such as memory.

6.4.2.2 Constrained Optimization

The problem defined in Figure 6.14 is a 0–1 integer optimization problem. The objective is neither concave nor convex as a function of the decision variable. It also is not continuous. Most of the constraints are nonlinear. For this complicated problem, conventional convex optimization approaches may not be applicable. Heuristic algorithms (e.g., simulated annealing algorithms and genetic algorithms) can be used to obtain suboptimal solutions.

One of the many challenges with the optimization is the approach used to address the constraints. First, the problem may not be feasible. Second, even if the problem is feasible, it takes time to find the feasible solution at each iteration of the searching algorithms, given the very large spanning space of the decision variables. To alleviate the problems with the feasibility, the constraints can be addressed by introducing penalties to the objective functions that can speed up the optimization process significantly. On the other hand, addressing the constraints in such a soft way should not affect the final results much since the inputs to the global controllers, including the workload demand and the power models, are only estimated or approximated, and the performance and budget constraints will be enforced by the local or feedback controllers in real time.

Another challenge with the optimization problem is the potential conflict between the constraints. For instance, enforcing the budget constraint will result in workload consolidation, which can lead to very high utilization of the servers and even performance loss due to full utilization of the servers. Setting different weights for the penalty functions corresponding to the priorities of the constraints can address this issue.

6.4.3 Case Study: Evaluation of the Integrated Management Solutions

6.4.3.1 Utilization Trace-Based Simulation

Ideally, the coordinated management architectures at the data center level should be evaluated in a real implementation. However, it could be difficult to do so for several reasons: (1) It is difficult to obtain access to a data center or a sufficiently large collection of machines. (2) Such a collection needs to be fully populated with relatively new servers with support for multiple power states. (3) All the individual controllers need to be set up and tuned. In addition to the effort needed, this allows only implementations specific to the idiosyncrasies of the systems considered. (4) It is difficult to set up the test bed with complex enterprise applications and exercise them to model real-world usage in real data centers. The alternate approach of using full-system simulation (e.g., M5, Simics, GEMS) suffers from the third and fourth drawbacks; also, simulation speeds and complexities of modeling clusters make this impractical. Given these challenges, several studies took recourse to trace-driven simulation for data center environments.

This approach uses real-world traces from actual enterprise deployments to drive individual server simulations. High-level models like those discussed previously are used to correlate resource utilization and the impact of changing specific actuators to system metrics like power and performance. This approach enables the workload behavior and system characteristics to be modeled expediently while allowing detailed evaluation of trade-offs at the policy and system parameter levels.

As one example, we show a set of evaluation results through trace-driven simulation for the architecture presented in Figure 6.6. Multiple metrics were used for the evaluation, including the aggregate power savings, the performance loss, and the power budget violations at the server, enclosure, and group levels. In the simulator, no queuing process was assumed when the aggregate demand of the workloads running on a server exceeded the capacity of the server. So, when the workload demand was increased, or the capacity of the server was reduced due to power capping, performance loss could happen as the excessive demand was not carried over.

The advantage of the trace-based simulation methodology is that it allows people to use actual utilization traces from real-world enterprises. Specifically, 180 traces were used that represent individual server utilization from nine different enterprise sites for several classes of individual and multitier workloads (database servers, web servers, e-commerce, remote desktop infrastructures, etc). To better study the variability in workloads, four mixes were defined: one incorporating all 180 workloads (180) and others focusing on specific mixes of 60 workloads (60L, 60M, 60H) with low, median, and high intensity of CPU loads. Most of the workload traces, as is common with most real-world deployments, showed relatively low utilization (15–50% in most cases). To better illustrate more resource-intensive workloads, "synthetic" workloads (60HH, 60HHH), which had much higher intensities than the real trace-based workloads, were created that stacked multiple workloads from the real-world traces to create higher utilization.

In this example, we studied two different kinds of enterprise systems: a low-power blade server, Blade A, and an entry-level 2U server, Server B with

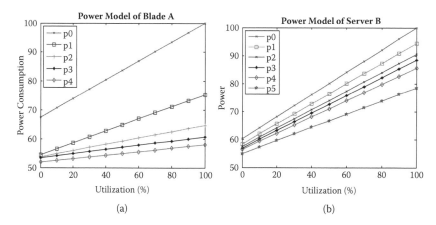

FIGURE 6.15   Power models for Server A and Server B for the evaluation of the integrated management solution.

performance-power models shown in Figure 6.15. The processor of Blade A had five P states, with frequencies of 1 GHz, 833 MHz, 700 MHz, 600 MHz, and 533 MHz. The processor of Server B had six P states, with frequencies of 2.6 GHz, 2.4 GHz, 2.2 GHz, 2.0 GHz, 1.8 GHz, and 1.0GHz. It was assumed that the baseline was virtualized. For VM migration, a precopied migration process was assumed, and the migration overhead was taken as 10% performance loss during the migration process.

A cluster of 180 servers was assumed to host the 180 workload. This was organized as six 20-blade enclosures and 60 individual servers. For the 60 workload evaluations, we assumed a cluster of 60 servers: two 20-blade enclosures and 20 individual servers. By making these mixes, we could have a cluster with both blades and individual servers that were representative of the data centers today.

The power budget levels had significant effects on the metrics. Three different kinds of power budget values were studied: (1) 20-15-10, representing group, enclosure, and local power budget caps that were respectively 20%, 15%, and 10% off from their maximum possible power consumption; (2) 25-20-15, representing caps that were 25%, 20%, and 15% off their maximum possible power consumption; and (3) 30-25-20 representing caps that were 30%, 25%, and 20% off their maximum. Note that the budgets were relative numbers with respect to the peak power consumption. With more stringent power budgets, the performance is expected to be compromised more significantly.

Besides the server models, the power models, and the budget levels, sensitivity analysis can be done with the alternative architecture. The sensitivity of the architecture was studied through a few alternatives, for instance, the control intervals. In the baseline, the control intervals of EC/LPC/EPC/GPC/GC were set to 1/5/25/50/500, respectively. Other alternatives include variants of the models with different idle power, P-state groups, coordination architectures, policies, and so on. But, in this chapter, we only discuss the basic results in the next section. Additional results are available [20].

6.4.3.2 Basic Results: Uncoordinated versus Coordinated Control

In the set of principal simulations, we used a system in which no controllers for power management were turned on as the baseline and compared two distinct solutions: (1) the *coordinated* architecture, as in Figure 6.6; (2) an *uncoordinated* solution, in which the individual power management solutions for power efficiency, power capping, and power optimization worked independently of one another. Figure 6.16 shows the results for both the coordinated and the uncoordinated solutions compared against the baseline results. Four configurations are included, representing two types of systems (Server A and Server B) and two sets of workloads (180 and 60HH). For each configuration, we present a family of four bars: three bars for power budget violations at the group, enclosure, and local levels and one bar for performance degradation. To visually illustrate the negative ramifications of budget violations and performance loss, we show these as negative numbers. Note that some of the bars are not visible since the budget violations were very low or even zero when compared with the baseline. For instance, the violations at the GC level were all zero, which means that the group power budgets were well maintained.

The top left graph in Figure 6.16 shows the results for the base 180 server configuration for Blade A. Compared to the baseline, the coordinated solution achieved a 64% reduction in power consumed (not graphed), translating to savings in electricity costs, with negligible (3%) performance degradation and (5%) power budget violations. Recall that this configuration had additional savings of 10%, 15%, and 20% in the peak power budgets at the local, enclosure, and group levels, respectively, which translate to capital savings for the cooling equipment. In comparison, the uncoordinated solution resulted in greater performance loss (12%) and higher power budget violations (7%).

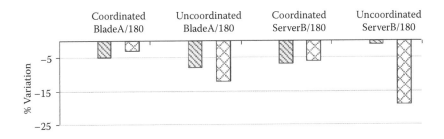

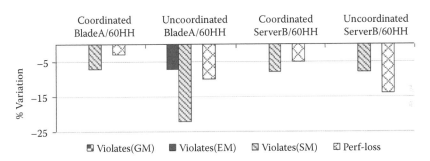

FIGURE 6.16 A comparison of an uncoordinated deployment and the co-ordinated solution for four different configurations. All results are normalized to a baseline where no controllers for power management are turned on. The two figures correspond to the two types of workload combination. There are four bars for each server/workload combination. The three left bars show violations in group, enclosure, and server power budgets. Some of them are invisible since the violation is zero; the last bar shows performance loss. In general, the uncoordinated architecture has higher performance degradation and power budget violations.

Figure 6.16 also illustrates the sensitivity to different system models. As discussed, Server B had six P states relatively uniformly clustered, but with a smaller range in power, compared to the five nonuniformly clustered, but higher-range, P states of Blade A. This typically manifested itself in reduced absolute power savings results for Server B compared to Blade A. It indicates that the range of power control was likely more important than the granularity of control for these configurations.

As discussed, in addition to the 180-workload configuration, we studied other workload sets with different levels of activity. The benefits from coordination were qualitatively similar for all classes of workloads. However, as one would expect, the actual power savings for the low-utilization workload relative to the baseline was higher than that for the high-utilization work-load, while the relative improvements over the uncoordinated solution were higher with higher utilizations.

6.5 CONCLUDING COMMENTS

In this chapter, we discussed how, given the growing challenge from power and cooling, future data centers will likely deploy multiple power management solutions at the same time, and federation of these solutions is desirable. It therefore becomes important to consider a solution that coordinates different power solutions across the various axes of the taxonomy. This chapter tried to address the key questions that arise in the context of such a cross-layer power management architecture. How should individual controllers interact with each other to ensure correctness, stability, and efficiency? In particular, how do we federate the individual controllers to be aware of one another but without requiring global knowledge of all the properties at each of the individual controllers? Furthermore, given the dynamism in future enterprise environments, how do we design the solution to respond to changes in the number and nature of controllers participating in the overall architecture and to changes in the nature of systems and applications deployed?

This work raised an additional set of interesting questions that we did not consider. Specifically, these questions pertain to the implications of such a unified solution on the design and deployment of individual power management solutions. Are all solutions equally important? Does the coordinated architecture allow for functionality of one controller to be simplified, or even subsumed in another controller, to enable an overall simpler design? Do the policies and mechanisms at the individual level need to be revisited in the context of their interactions with other controllers? How sensitive are the answers to these questions to the nature of applications and systems considered?

In addition, it is important to consider mechanisms for information exchange and policy brokering in the context of such a federated cross-layer power management solution. Specifically, aggregation and coordination of data across the infrastructure layers are required for integrated operation. To enable this, we need a corresponding solution to loosely couple different management algorithms and facilitate monitoring and management coordination in data centers. One solution that we have considered is comprised of registry and proxy mechanisms that provide unified monitoring and actuation across platform and virtualization domains and coordinators that provide policy execution for better VM placement and run-time management, including a formal approach to ensure system stability from inefficient management actions. The work [21] presents more details on this approach.

This area, however, is still nascent, and we believe that the next few years will see much more progress in this field. Although this chapter focused on power management, it is representative of a broad class of problems typified by "intersecting control loops." We expect the principles and discussion outlined in this chapter to generalize to the broader resource management domain. Overall, as the complexity of management continues to increase, with multiple players at multiple levels optimizing for multiple objectives, approaches like the ones discussed in this chapter that focus on coordination across these multiple levels are likely to be a critical part of future enterprise architectures.

ACKNOWLEDGMENT

We would like to thank Ramya Raghavendra, Xiaoyun Zhu, VanishTalwar, Cullen Bash, Niraj Tolia, and Manish Marwah for their contributions to part of the content in this chapter.

REFERENCES

1. ACPI: Advanced configuration and power interface home page. http://www.acpi.info.
2. L. A. Barroso and U. Holzle. The case for energy-proportional computing. *IEEE Computer*, 40(12):33–37, 2007.
3. C. D. Patel, C. E. Bash, R. Sharma, M. Beitelmam, and R. J. Friedrich. Smart cooling of data centers. In *Proceedings of IPACK'03*, Kauai, Hawaii, July 2003.
4. S. Greenberg, E. Mills, B. Tschudi, P. Rumsey, and B. Myatt. Best practices for data centers: Results from benchmarking 22 data centers. In *Proceedings of the 2006 ACEEE Summer Study on Energy Efficiency in Buildings*, Pacific Grove, CA, August 2006.
5. The Green Grid home page. http://www.thegreengrid.com.
6. C. E. Bash, C. D. Patel, and R. K. Sharma. Dynamic thermal management of air cooled data centers. In *Proceedings of ITHERM*, pages 445–452, San Diego, CA, May 2006.

7. T. Heath, A. P. Centeno, P. George, L. Ramos, Y. Jaluria, and R. Bianchini. Mercury and freon: Temperature emulation and management for server systems. In *Proceedings of ASPLOS*, pages 106–116, San Jose, CA, October 2006.

8. C. Patel, R. Sharma, C. Bash, and S. Graupner. Energy aware grid: Global workload placement based on energy efficiency. In *Proceedings of IMECE 2003*, Washington, DC, November 2003.

9. J. Moore, J. Chase, P. Ranganathan, and R. Sharma. Making scheduling cool: Temperature-aware workload placement in data centers. In *Proceedings of the Annual Conference on USENIX Annual Technical Conference*, pages 61–75, April 2005.

10. C. Bash and G. Forman. Cool job allocation: Measuring the power savings of placing jobs at cooling-efficient locations in the data center. In *USENIX Annual Technical Conference*, pages 363–368, Santa Clara, CA, June 2007.

11. R. Ayoub, S. Shari, and T. S. Rosing. Gentlecool: Cooling aware proactive workload scheduling in multi-machine systems. In *Proceedings of DATE 2010*, Dresden, Germany, March 2010.

12. Q. Tang, S. K. S. Gupta, and G. Varsamopoulos. Thermalaware task scheduling for data centers through minimizing heat recirculation. In *Proceedings of the 2007 IEEE International Conference on Cluster Computing (CLUSTER '07)*, pages 129–138, Austin, TX, September 2007.

13. L. Wang, A. J. Younge, T. R. Furlani, G. von Laszewski, J. Dayal, and X. He. Towards thermal aware workload scheduling in a data center. In *Proceedings of the 10th International Symposium on Pervasive Systems, Algorithms and Networks (I-SPAN 2009)*, Kao-Hsiung, Taiwan, December 2009.

14. P. Ranganathan, P. Leech, D. E. Irwin, and J. S. Chase. Ensemble-level power management for dense blade servers. In *33rd International Symposium on Computer Architecture (ISCA 2006)*, pages 66–77, Boston, June 2006.

15. C. Lefurgy, X. Wang, and M. Ware. Power capping: A prelude to power shifting. *Cluster Computing*, 11(2):183–195, 2008.

16. A. Qureshi, H. Balakrishnan, J. Guttag, B. Maggs, and R. Weber. Cutting the electric bill for Internet-scale systems. In *Proceedings of SIGCOMM 2009*, pages 123–134, Barcelona, August 2009.

17. C. Stewart, T. Kelly, and A. Zhang. Exploiting nonstationarity for performance prediction. In *Eurosys 2007*, Lisbon, March 2007.

18. Z. Wang, Y. Chen, D. Gmach, S. Singhal, B. J. Watson, W. Rivera, X. Zhu, and C. D. Hyser. AppRAISE: Application-Level Performance Management in Virtualized Server Environments. In *IEEE Transactions on Network and Service Management*, 6(4):240–254, December 2009.

19. N. Tolia, Z. Wang, P. Ranganathan, C. Bash, M. Marwah, and X. Zhu. Unified power and cooling management in server enclosures. In *Proceedings of the ASME/Pacific Rim Electronic Packaging Technical Conference and Exhibition (InterPACK '09)*, San Francisco, July 2009.

20. R. Raghavendra, P. Ranganathan, V. Talwar, Z. Wang, and X. Zhu. No power struggles: Coordinated multi-level power management for the data center. In *Proceedings of ASPLOS*, pages 48–59, Seattle, WA, March 2008.

21. S. Kumar, V. Talwar, P. Ranganathan, R. Nathuji, and K. Schwan. M-channels and M-brokers: Coordinated management in virtualized systems. In *Workshop on Managed Multi-Core Systems (MMCS)*, Boston, MA, June 2008.

CHAPTER 7

Energy-Efficient Virtualized Systems

Ripal Nathuji and Karsten Schwan

CONTENTS

7.1   INTRODUCTION

The semiconductor industry has shifted toward multicore system architectures to continue harnessing the resources made available by Moore's law, while avoiding the power and thermal bottlenecks associated with high-performance single-processor architectures. The effects of this transition are already apparent in data center environments, where commodity servers often consist of dual-package configurations that allow on the order of ten processors to be provisioned on a single platform. The increasing density of computational elements is accompanied with pressure on other platform components, such as memory and I/O (input/output). The net result is the profileration of dense server platform and rack configurations. Although these systems offer significant advantages for scale-up workloads, many enterprise applications are not able to fully exploit such architectures. The reasons for this include the inability for the associated software to scale to a larger number of processors, as well as the varying loads exhibited by web service applications. Unmitigated, the inability for applications to consistently and effectively make use of multicore server platforms can result in stranded data center resources, which can equate to significant cost inefficiencies. What is needed, then, is a means to manage resources in a fluid fashion, thereby achieving an elastic data center where servers make up a fungible resource pool that can dynamically be provisioned to applications in an on-demand manner. It is becoming increasingly clear that virtualization technologies can help meet this goal.

System virtualization enables workloads to be easily consolidated onto, and migrated between, physical servers. This can enable, for example, the consolidation of scale-out applications onto multicore servers, as well as the redeployment of workloads as a function of load. Based on these management capabilities, virtualization has quickly become a fundamental underlying technology for data centers. For example, emerging cloud systems such as Amazon's Elastic Compute Cloud (EC2) [1] and Microsoft's Windows Azure Platform [2] leverage virtualization to improve the manageability of hosted applications, as well as for the security and isolation benefits that they provide [3]. Indeed, a system virtualization layer is often a de facto component of the stack in modern enterprise deployments.

While the presence of a virtualization layer helps disassociate applications from the physical resources on which they run, there remain challenges and opportunities in extending virtualization architectures for improved system management. Specifically, given the critical importance of active energy management in data centers, it is important to investigate how additional interfaces and mechanisms can be introduced to better manage power in virtualized systems.

Realizing the importance of energy management and virtualization, in this chapter we discuss the nexus of these two problem domains. We begin by providing a brief overview of how virtualization is an essential technology for enabling dynamic resource provisioning across physical server boundaries, serving as a first step towards realizing resource fungibility. We describe how current virtualization systems remain lacking when attempting to perform active power management while still supporting the quality-of-service (QoS) requirements of virtual machines (VMs). Based on these discussions, we then describe avenues for extending existing virtualization architectures to better support energy efficiency while managing systems in a VM-aware manner. To illustrate the benefits of such strategies, we highlight examples of these approaches based on our own research. Finally, we conclude with an outline of related work in the field of resource management of VMs, as well as thoughts on future challenges and opportunities for improving the energy efficiency and manageability of virtualized data centers.

7.2 VIRTUALIZATION AND POWER MANAGEMENT

In this section, we discuss the role of virtualization in modern data centers and provide an architectural overview of the virtualization technologies widely deployed today. We then describe how the addition of virtualization to server environments affects the power management strategies developed for native, nonvirtualized servers. Specifically, we highlight how, without additional support, simply carrying forward methods from nonvirtualized systems does not achieve desired management goals. Finally, we identify opportunities and directions for extending virtualization technologies to overcome these limitations.

7.2.1 Benefits of Virtualized Data Center Deployments

To more easily motivate the role of virtualization in large-scale data centers, we begin by discussing the benefits that virtualization provides over managing physical infrastructures. In nonvirtualized environments, the decision

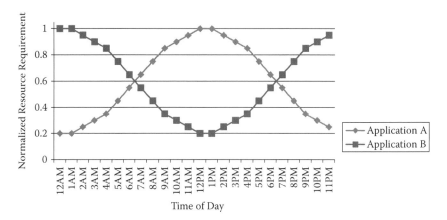

FIGURE 7.1   Example of resource requirements with diurnal load behavior.

of mapping workloads to physical servers is predominantly a static optimization. For example, an administrator may allocate resources to an application based on the capacity required to meet a service-level agreement (SLA) for an application at peak load. Whenever the application experiences reduced load, or can be allocated fewer resources while still meeting SLAs, resources that were reserved for it go unused. Given that data centers typically host multiple applications simultaneously, this type of "resource siloing" can result in significant underutilized capacity. Specifically, provisioning in this manner requires sufficient computational resources to meet the *sum of peak* loads across the applications hosted by the data center.

To illustrate the potential resource inefficiency of silos, Figure 7.1 provides an example of applications with varying resource demands deployed in a data center. Both applications exhibit diurnal resource usage characteristics based on load requirements. However, they are out of phase with each other, with one application experiencing high demand during the day, the other at night. With the resource silos that occur from traditional physical infrastructures, each application will be provisioned with a separate pool of resources that allow it to meet peak load. This results in 40% of data center capacity being left unused at any given point in time because in reality neither application uses all of its resources at the same time. Even if the idle resources can be placed into very low power states [4], the costs of the hardware, as well as the costs associated with the power and cooling infrastructures they require, are not recovered with productive computations. Instead, if one can provision resources equal to the *peak of sum* resource usages between the two applications, we can reduce the overall capacity required by 40%. To do this, however, we must have a means of treating data center capacity as a fungible pool of resources that can be dynamically allocated to applications at run time based on demand.

Attaining true resource fungibility at data center scale is a challenging problem. Virtualization technologies provide a significant first step toward realizing this vision. Specifically, virtualization allows data center management systems to dynamically adapt resource allocations to guest VMs and move workloads between physical servers when necessary [5]. While similar functionality can be achieved without virtualization support, doing so requires substantial changes to existing application and operating system software. Therefore, by incorporating virtualization into data centers, we start to achieve the agility necessary for resource fungibility while preserving compatibility and support for existing software stacks.

Observing the expanding demand for virtualization, multiple solutions have been developed and received significant adoption, including the open source Xen system [6], VMware's ESX hypervisor [7], and Microsoft's Hyper-V software [8]. We next present the relevant architectural components of virtualization solutions that we assume in our discussions in the remainder of this chapter.

7.2.2 Virtualization Architectural Overview

In this chapter, we concern ourselves with type 1 hypervisors that run directly on server hardware, making use of hardware extensions to enable efficient processor virtualization. This is in contrast to type 2 hypervisors, which are hosted on top of an existing operating system environment, examples of which include Microsoft Virtual PC [9] and QEMU [10]. By executing directly on the physical platform, type 1 hypervisors have the benefit of precisely controlling resources and managing low-level hardware, including power management states. Both Xen and Hyper-V adopt a lightweight hypervisor that performs scheduling and memory management but delegates higher-level operations for VM-level management, including creation, destruction, migration, and resource reprovisioning, to a management partition. The management partition is itself a VM, albeit with privileges that allow it to perform the required functionality.

Guest VMs that are hosted on top of hypervisors can be separated into those that are paravirtualized with modifications made explicitly for execution in virtualized systems and those that are not modified for virtualized execution. While supporting the latter is required to enable any guest operating system and application stack to execute on a virtualized server,

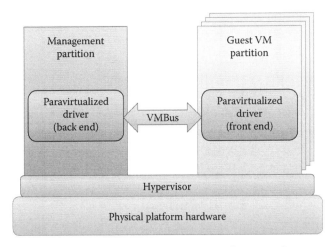

FIGURE 7.2   Illustration of key virtualization architectural components.

introducing paravirtualization can have significant benefits. A clear example of this can be found in the context of device virtualization. Without paravirtualization support, device virtualization can exhibit poor performance due to emulation overheads. To help remove such bottlenecks, virtualization systems often employ paravirtualized drivers, where a front-end component executing in the guest VM communicates with a back-end component running in the management partition. This communication is performed over a virtual bus abstraction (e.g., XenBus in Xen and VMBus for Hyper-V). While there will always be need for supporting legacy guests that do not incorporate these types of changes, paravirtualization introduces interesting avenues for improving interactions between VMs and the management partition.

Figure 7.2 provides an illustrative overview of the key components of virtualization solutions outlined. As described, a hypervisor that runs directly on hardware is coupled with a management partition. The combination of hypervisor and management partition supports the execution of guest VMs, including, where applicable, providing paravirtualized service components to match up with synthetic devices used by VMs. What is missing from this picture, however, is what infrastructure support is necessary for effective power management of virtualized systems. To address this question, we next consider fundamental power management goals in nonvirtualized systems and how the introduction of virtualization might place additional requirements on the system to continue to meet them.

7.2.3 Power Management Requirements for Virtualized Systems

Managing energy in server environments is often geared toward one of two goals. The first is to optimize the active power consumption of physical servers while maintaining acceptable QoS for the hosted application, with a typical approach being to reduce processor voltage/frequency states based on processor utilizations. The second goal for server power management is to treat power as a constraint via server power budgets. Such power capping can be used to provision additional physical servers within the limited power capacity of a data center [11]. The trade-off here is based on the same sum-of-peak versus peak-of-sum provisioning that was discussed in Section 7.2.1 in the context of stranded resources. Here, the goal is to prevent stranding of power [12] by provisioning power based on an average or typical power consumption instead of the peak usage per server. This type of power overcommitting can be performed safely as long as there are power-capping mechanisms available to enforce power consumption limits when necessary. Both of these goals affect both platform-level and distributed management policies.

At the platform level, whether optimizing power consumption during the execution of an application or enforcing power caps on a server, the system must carefully balance the requirements of the application with the performance/power characteristics of underlying power management states. For example, considering dynamic voltage and frequency scaling (DVFS) of processors, policies may carefully toggle processor performance states to optimize energy consumption while meeting QoS in terms of real-time execution constraints [13, 14]. In the case of physical servers, there is an implicit assignment of a single application to a platform, thereby allowing policies that are tuned for the application and hardware to drive the underlying power management performed by the operating system. In virtualized systems, however, there is no direct way to map this type of application-specific policy feedback since each physical server hosts multiple, possibly heterogeneous, applications with varying power/performance trade-offs. Moreover, as previously mentioned, the management of physical hardware states is controlled by the hypervisor and management partition, further preventing guest VMs from directly toggling power states. *What is required, then, is the extension of virtualization infrastructures to better enable guest VMs to interoperate with the virtualization stack to perform power management decisions.* In Section 7.3, we provide examples of these types of virtualization enhancements based on our own previous work. Specifically, we discuss the inclusion of VM feedback, either through the virtualization of legacy power management interfaces or through paravirtualized management interfaces.

Managing power across distributed servers requires coordination across the many localized management entities on different platforms. Consider power capping as a concrete example; power caps typically need to be enforced at some aggregate level (e.g., rack or blade chassis [15]), but the joint budget is often controlled by allocating platform-level budgets that are then enforced by local controllers [16]. For efficient distribution of power resources, this requires some level of interaction and coordination between servers so that, for example, when one system is not using its allocated power capacity, that capacity can be provisioned to others. In the case of virtualized servers, each physical server hosts a set of VMs, adding an additional level of hierarchy to the system. Hence, where before distributed policies may have intelligently managed resources between physical machines, there must now be some additional level of awareness that allows for management across both physical platforms and the virtual instances they host. *Realizing this goal requires coordination across the virtualization layers controlling each of the distributed physical platforms.* We demonstrate these ideas, and the benefits they can provide, in the context of VM aware power budgeting for distributed servers in Section 7.4.

7.3 PLATFORM ENHANCEMENTS FOR ENERGY-AWARE VM MANAGEMENT

The increasing emphasis on energy efficiency has resulted in the deployment of power management policies that are tuned for specific workloads or classes of applications. For example, operating systems, including Linux and Microsoft Windows, incorporate utilization-driven policies that manage processor voltage/frequency states. The parameters that govern the behavior of these policies can be fine-tuned for individual applications to improve power savings while meeting performance requirements. The existence and deployment of these policies help meet application-specific QoS constraints when performing active power management. Therefore, it is highly desirable to continue leveraging these policies under virtualized execution.

Efforts toward performing power management using VM-based policies must simultaneously meet constraints of isolation and independence. Isolation refers to the fact that power management actions performed on behalf of one VM should not adversely affect another, cohosted, application. Similarly, any attempt to make use of VM-specific policies must maintain the independent manageability expected by guests. To meet these design requirements, we pursue an approach where VM management policies convey power/performance trade-offs to the management partition. The information is then used as feedback to authoritative policies in the management partition, which eventually modify resource allocations and power states of physical resources. In this section, we describe two examples that illustrate this idea. First, we consider leveraging existing power management interfaces through the VirtualPower [17] system to obtain VM input. We then relax the constraints around relying on legacy interfaces by considering the integration of paravirtualized feedback for QoS-aware server power budgeting [18].

7.3.1   Coordinated VM Power Management with VirtualPower

7.3.1.1   *VirtualPower Architectural Overview*

The VirtualPower system is based on three major architectural components. These consist of VirtualPower management (VPM) states, VPM channels, and VPM mechanisms. We describe each of these three in turn to provide an architectural overview of VirtualPower.

7.3.1.1.1   VPM States

Typically, platforms expose hardware-supported power management states to system software and applications via ACPI [19]. VirtualPower exploits this existing legacy interface to convey a set of "soft" power management modes, known as VPM states, to VMs. In this manner, VPM states allow the VirtualPower system to add a level of indirection between guest power management policies attempting to modify physical power states and the actual physical system configuration. An added benefit of VPM states is that they provide a consistent view of manageability to guest VMs. Observing that data centers are often comprised of heterogeneous servers with different power manageability characteristics [20], VPM states allow VMs to remain agnostic of these underlying changes as they are migrated between heterogeneous hardware.

The inclusion of VPM states allows VirtualPower to enable guest-level power management policies while meeting the desired qualities of isolation and independence mentioned. Specifically, modifications made to the value of VPM states do not themselves directly affect changes to physical performance states. For example, a VM policy change to a locally exposed processor frequency (e.g., P-state in ACPI) only affects the soft state associated with the corresponding guest, with no visible changes to other hosted operating systems. For this reason, independence is achieved as well because actuations made by the guest are not observable by other VMs. Hence, there is no creation of dependencies between individual VM policies that might require making guest policies aware of each other. Of course, as described thus far, VM-level policies simply make changes to exposed VPM states without actually modifying system behavior. To affect system changes, VirtualPower communicates changes made by VM policies to the management partition using VPM channels.

7.3.1.1.2   VPM Channels

As described in Section 7.2.2, explicit communications between guest VMs and the management partition are usually performed over a virtual bus abstraction such as VMBus for Hyper-V. In the case of VirtualPower, however, the policy component in the guest is not aware that it is running in a virtualized context and actuates VPM states using standard mechanisms such as writes to model-specific registers (MSRs). When the policies perform these privileged operations, a fault occurs transferring control to the hypervisor. With VirtualPower, the hypervisor infers when these faults occur due to power management actions made by guest policies. Based on this knowledge, the hypervisor constructs a VPM event object that includes information such as the current time stamp, the VPM state that the guest attempted to set, and the physical performance state being used by the system for the VM when the event was detected. These events are stored in cyclic buffers by the hypervisor until they are retrieved by policies running in the management partition. Specifically, the management partition can use hypercalls to both poll for and acquire outstanding VPM events. In addition, the hypervisor makes available an event channel that the management partition can subscribe to if it wishes to be notified when events occur. The combination of fault-generated VPM events and event notification interfaces in effect provides a communication medium between VM-based policies and the management partition that we refer to as VPM channels.

7.3.1.1.3   VPM Mechanisms

The combination of VPM states and VPM channels allows the management partition to ascertain power/performance trade-offs from the perspective

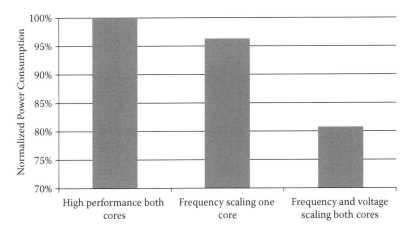

FIGURE 7.3 Limited hardware power savings resulting from dependencies between managed VMs on a dual-core processor.

of each guest VM that is hosted on a server platform. The set of actions that policies in the management partition, VPM rules, can subsequently perform to affect application performance and power consumption are termed *VPM mechanisms*. The first mechanism that can be considered is of course the use of underlying hardware states. When using these states, however, the management partition must make sure that it does not violate the isolation principles provided by VPM states. VirtualPower enables this by implementing the notion of shadow VPM states. Shadow VPM states consist of the actual configuration that the hypervisor will use when provisioning virtual resources at a VM-specified VM state. Taking processor voltage/frequency states as an example, the management partition may determine that each soft VPM state should result in a distinct hardware-supported frequency state to be set on the physical processor on which a virtual processor is running. Isolation is guaranteed here since physical states are modified as virtual processors are context switched. Note that the second requirement identified, independence, still exists since guest policies may still operate in an independent manner. However, there may exist dependencies that effect power savings that the management partiton VPM rules may take into account when utilizing VPM mechanisms. For example, a limitation of depending exclusively on hardware management states that motivates the inclusion of additional VPM mechanisms is indicated in Figure 7.3.

The power consumption of a processor is proportional to the product of frequency and voltage squared. Therefore, voltage scaling is required to obtain significant power savings. However, often there may be fewer voltage rails than available processors, and the effective voltage is limited by the fastest running processor. When the shadow VPM states assigned to the virtual processor of distinct VMs are different, the ability of VPM rules to utilize hardware states for power savings may be limited. As illustrated in Figure 7.3, frequency scaling one processor on a dual-core package independently of another provides nearly 5% in system-level power savings. However, scaling both processors enables voltage scaling as well, dramatically increasing platform power reduction by nearly 20%. Based on this observation, VirtualPower extends additional VPM mechanisms for VPM rules to utilize. One of these, soft scaling, utilizes resource scheduling to reduce the effective performance experienced by a virtual resource by reducing the amount of physical resource it obtains.

Using processors as an example again, VPM rules can "cap" a virtual processor such that the hypervisor imposes utilization limits on it [18]. Soft scaling can provide power benefits because it duty cycles the underlying hardware, allowing it to enter lower-power idle states. Indeed, with the increased emphasis on idle power management, soft scaling can help hardware attain energy-saving states that improve power consumption beyond the hardware capabilities available. As an illustrative example, Figure 7.4 compares scaling the performance of a virtual processor by duty cycling the processor during VM execution using hardware throttle states versus hypervisor scheduling. We observe that at lower performance states, the

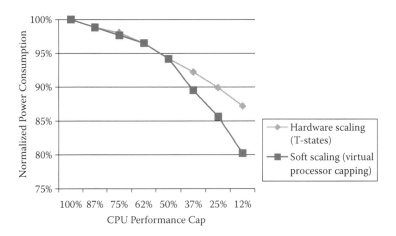

FIGURE 7.4   Power benefits of low-power hardware states achieved through the soft-scaling VPM mechanism.

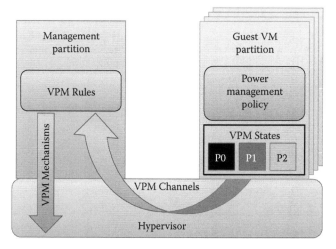

FIGURE 7.5   Overview of the VirtualPower management architecture.

higher time constants used by soft scaling enable low-power C states [19] on the processor, which provide a better power profile than hardware-based throttling. In general, we find that soft scaling can be an effective and flexible mechanism that can be exploited by VPM rules when hardware mechanisms either are limited or dependencies between VMs render them unproductive. A final VPM mechanism that should be mentioned is the use of consolidation. Here, consolidation may be the collocation of virtual resources within a physical platform, as well as the more common use of VM migration. In either case, the consolidation of virtual resources, possibly in conjunction with soft scaling, enables idle states that VPM rules can use to reduce power consumption.

7.3.1.1.4   Summary

Figure 7.5 illustrates the overall VirtualPower architecture based on the components described in this section. The system exposes a set of soft VPM states to each guest. The internal guest policies actuate on these virtualized performance states based on the application-specific performance criteria and the actual experienced levels of QoS. Information regarding these actuations are conveyed to VPM rules in the management partition. Finally, the VPM rules utilize the information to appropriately exercise VPM mechanisms, which in return affect system power consumption and the performance experienced by VMs, thereby completeting the feedback loop. An interesting artifact of this design is that the feedback signals back to the guest VMs are implicit since the guest policies are not paravirtualized and do not communicate directly with the management partition. However, under the assumption that they are driven by application performance, actions performed by VPM rules affect guest policies. We next present some illustrative experimental results that demonstrate the efficacy of the VirtualPower system.

7.3.1.2 Experimental Results

As a first set of experiments, we exercised a VirtualPower-enabled system with VMs running a simple CPU-bound transactional workload. We considered varying performance requirements for this application, where the SLA may be relaxed at times. For example, for transaction-oriented back-end systems, the overall response time of transactions may be dominated by queuing delays when a transient large batch of requests arrives. This can allow the system to trade off power consumption for processing throughput. Similarly, a transaction-oriented service may guarantee results within some time frame, allowing it to experience periods of reduced performance as long as some minimal prescribed processing rate is maintained. Based on this application model, we implemented VM policies that monitor active transaction processing rates and, based on a comparison to a reference SLA rate, modify VPM states. Figure 7.6 contains the power consumption of a multicore platform hosting transactional VMs.

Figure 7.6a illustrates the power implications that a VirtualPower enabled system can have using VM input when a single VM is deployed in the platform. Specifically, we see that when the application requires less-than-peak processing rates, the VPM rules in the management partition can make use of VPM mechanisms to reduce power consumption of the server. Without the VPM events communicated via the VPM channel, however, the policies could not determine the workload-specific trade-offs. Even if one were to implement utilization-based DVFS policies in the management partitition, as previously mentioned, the transactional workload is CPU bound. Hence, without the workload-specific insights obtained via feedback from VM policies, such a policy would conservatively have to maintain peak performance. Instead, we observed power reductions of up to approximately 30% based on VM input with VirtualPower. Figure 7.6b, consider, the case of two VMs executing transactional applications that are consolidated onto the platform. We observe that as the performance requirements are reduced for one, or both, VMs, the system is able to effectively manage

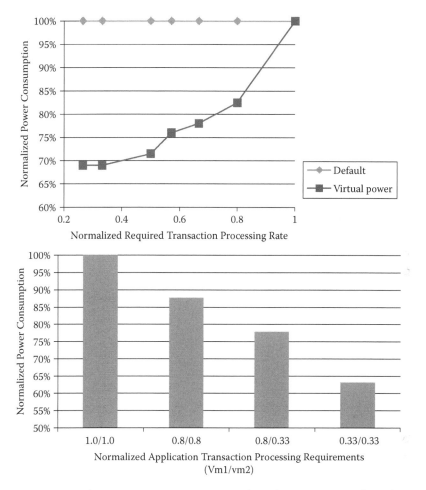

FIGURE 7.6 Reducing power consumption based on VM-specific policies with VirtualPower.

power consumption using a combination of VPM mechanisms, including both hard and soft scaling.

A second set of results highlights the ability of VirtualPower-enabled power management to address applications that have SLAs based on metrics with rich semantics. We modified the Nutch search engine application [21] to incorporate a quality model for the responses it generates to queries. Specifically, we extended the application to use a quality-of-information (QoI) metric that is associated with each search response. The modified server responds to requests by attempting to meet the quality requirements specified by a query within a latency bound. If the latency period is reached,

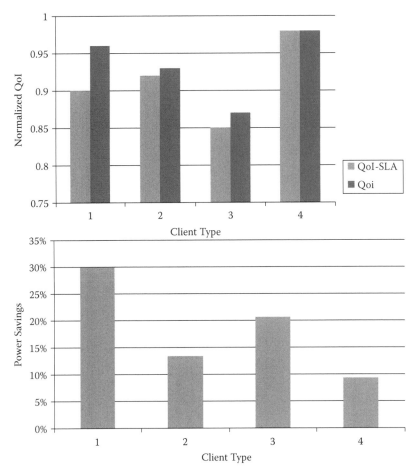

FIGURE 7.7 Managing power with QoI performance metrics.

the server responds with the results it was able to collect up to that point. The QoI of a response is then calculated as a function of the number of results required in the response and the number the server was actually able to return. We exercise the server with different client types that have varying QoI and latency requirements. The VirtualPower system manages the platform based on input from guest VM policies that attempt to manage virtualized management states as a function of measured and desired QoI. The ensuing results are summarized in Figure 7.7.

We observe in Figure 7.7a that, across all of the client types, the VirtualPower-enabled system is able to meet, or exceed, the QoI prescribed by the SLA. At the same time, the system is able to reduce power consumption by up to 30%. These power savings are attained while meeting SLAs, in spite of the fact that the management partition is completely unaware of the application QoS semantics of the hosted applications. These results help demonstrate the benefits of leveraging power management policies that execute VMs in a transparent manner by virtualizing existing management interfaces. Next, we consider the use of paravirtualized communications between the management partition and VMs for power management.

7.3.2 Paravirtualized Management Interfaces for Platform Power Budgeting

As previously mentioned, power capping in server environments is a common tool used to more aggressively provision servers within a given power capacity [11]. The ability to limit power consumption may also be used to manage thermal characteristics within a data center [22]. Power-capping mechanisms can be integrated into hardware [15, 16]. However, for virtualized systems, a key drawback of these approaches is that they throttle resources uniformly across the platform, affecting the performance of all VMs that are hosted on the platform equally. To demonstrate the use of paravirtualized communications for more effective power management of virtualized servers, we briefly consider an alternative approach by which software techniques utilize soft scaling to reduce power consumption in a QoS-aware manner.

7.3.2.1 QoS Feedback with Congestion Pricing

Enforcing power budgeting requires the system to limit the amount of active resources in the system or the performance levels at which they run. For example, the hardware and soft-scaling VPM mechanisms described can be used to limit power consumption. For simplicity, we limit ourselves to soft scaling in the following discussions. One method to enforce budgets using soft scaling is for the system to monitor power consumption from the hardware and use it as feedback to determine a system-wide resource cap r_{cap} that meets the specified power limit. This can be performed using a simple PI-, PD-, or PID-based controller (for instance [16, 18] or more sophisticated controllers. In general, some feedback control can be implemented to determine a system-wide cap that can, for example, be divided by the number of VMs to determine resource allotments for each workload, enforced via soft scaling. However, as previously mentioned, our goal is to support QoS-aware resource provisioning under power budgets.

Supporting differentiated services under power budgets requires some method to account for VM-specific QoS trade-offs. One method of achieving this would be to allow guests to specify some type of utility function that describes the trade-off between resource allocations and the costs associated with them (e.g., power). While this lends itself to centralized solutions that can optimize a system based on knowledge of the individual applications, a distributed solution provides better scalability and removes the requirement for explicit utility functions that may be difficult to realize in practice. Instead, we consider the use of paravirtualized interfaces that allow guest VMs to provide more meaningful feedback than what can be expressed through legacy interfaces such as ACPI. In particular, based on previous work, we employ a decentralized framework that uses congestion pricing [23–25] as a feedback mechanism for power budgeting of virtualized servers.

Congestion pricing allows the governing policies in the management partition to signal application VMs when the system is resource constrained using the notion of *shadow prices*. Applications then react to fluctuations in prices by altering the amount of resources they request. Here, the shadow price p_i is driven by the "costs" associated with provisioning a VM with some resource allocation r_i (e.g., based on soft scaling). Therefore, p_i is proportional to the resources allocated to the guest as well as the contribution to system congestion caused by the allocation. The guest provides feedback to the management partition by changing its bid B_i on the amount of resources it desires. In particular, the resource bid by a guest is a function of the shadow price p_i as well as the willingness of the guest to pay w_i. The willingness to pay value is VM specific and reflects the QoS priorities of the application that is running in the guest. Hence, the combination of shadow prices and resource bids makes up the paravirtualized communications between the management partition and guest VMs, while the values of w_i drive the bidding policies of each VM. The key to coordinating between the system resource manager in the management partition and the independent VM policies, then, is how the shadow prices are calculated in the system.

The system resource manager uses the set of bids received from guests to periodically update resource allocations and redefine shadow prices. As part of this, the manager calculates the "fair" virtual processor cap (e.g., soft scaling) F in the system based on the r_{cap} provided by a power budget controller implemented using a feedback loop as described. A fundamental constraint of the resource manager is that it must assign a cap of at least

F to a VM unless $B_i < F$, in which case a cap as low as the VM bid is sufficient. If a set of VMs is provided caps below F, it frees up resources to overprovision guests that are bidding above F. Therefore, at any given time, there are two types of costs in the system: an overage cost C_{over} based on guests that obtain a cap less than F and an opportunity cost C_{opp} based on VMs that have bids B_i greater than the received cap r_i. These costs are summarized by Equations 7.1 and 7.2.

$$C_{over} = \sum F - r_i \quad (\forall \, r_i < F) \tag{7.1}$$

$$C_{opp} = \sum B_i - r_i \quad (\forall \, r_i < B_i) \tag{7.2}$$

The resource manager uses an algorithm to update allocations based on VM bids B_i and then calculates current shadow prices using C_{over} and C_{opp}. In particular, based on F, the resource manager calculates the sum of all bids above F, B_{over}, and all other bids, B_{under}. These are then used to derive cap values as described in Algorithm 1. The algorithm uses the fair cap if either B_{over} or B_{under} is zero. Otherwise, it distributes the system-level cap r_{cap} to overprovision performance-sensitive VMs that have bid over F up to their respective bids while meeting at least the bid for all other VMs.

Algorithm 1 QoS-Aware VM Resource Allocations under Power Budgets

1: **if** $((B_{under} == 0)$ OR $(B_{over} == 0))$ **then**
2:    Set cap to $F \, \forall$ VMs
3: **else**
4:    **if** B_{under} allows B_{over} to be met within r_{cap} **then**
5:       $\forall$ VMs with $B_i > F$, set cap to bid
6:       $\forall$ VMs with $B_i \le F$, set cap to bid. Distribute remaining cap under r_{cap} to these VMs in proportion to B_i
7:    **else**
8:       $\forall$ VMs with $B_i \le F$, set cap to bid
9:       $\forall$ VMs with $B_i > F$, set cap to F. Distribute remaining cap under r_{cap} to these VMs in proportion to B_i
10:    **end if**
11: **end if**

Once determined, the resource manager uses the new virtual processor caps r_i to calculate costs and new shadow prices p_i. The shadow prices are communicated to VM agents using paravirtualized interfaces, after which the VM policies respond with new bids B_i. Specifically, shadow prices for

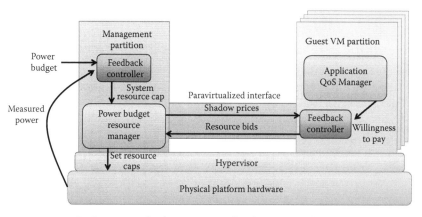

FIGURE 7.8 QoS-aware platform power-budgeting architecture with par-avirtualized VM feedback.

VMs with r_i greater than F are set to $\kappa_{over} C_{over} r_i$, and all others are charged $\kappa_{opp} C_{opp} r_i$, where κ_{over} and κ_{opp} are configurable parameters in the system. In terms of bid responses, since the w_i of a guest reflects its current QoS trade-offs, the goal of a VM during constrained periods is to place bids in a manner such that the shadow price p_i it is charged is approximately equal to its w_i. One method to achieve this is again to use a simple PID-based feedback controller in each guest that is driven by the error signal $(w_i - p_i)$ to adjust bids B_i. Overall, we achieve a control system that distributes control decisions across the management partition and guest VMs based on paravirtualized communications. Figure 7.8 summarizes the overall system. We next present some experimental results based on an implementation of this design.

7.3.2.2 Experimental Results

Figure 7.9 illustrates the use of soft scaling as a mechanism to cap power consumption of a server. In the experiment, the platform moves from an initial unconstrained state through phases where power is constrained to 98% and 94%. The figure provides the power samples at each 1-s interval, as well as a data curve based on averaging power values across a 5-s sliding window. We observe that, as expected with a feedback approach, the power deviates from the budget slightly at times. However, the controller is successfully able to converge to the power budget as it is changed, and adhere to it, thus validating the use of feedback control with soft scaling for power budgeting.

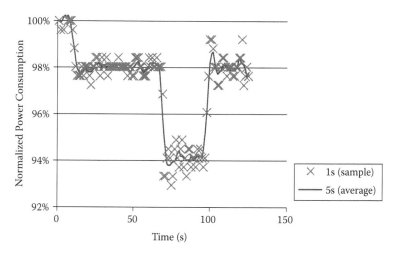

FIGURE 7.9   Efficacy of soft scaling for power budgeting.

Our power-budgeting system implements power budgets by varying resource allocations to VMs based on their conveyed QoS requirements. In Figure 7.10, we illustrate how the system manages resources as the QoS requirements of VMs, via the willingess-to-pay values (w_i), are varied. In the experiment, VM1 starts out with a higher value of w_i than VM2, and after some time, the VMs change their QoS so that the equivalent values are swapped. The results illustrate the ability of the resource manager to react to and improve capping based on relative workload demands. For example, during its demanding period, the system is able to provision VM1 with additional resources compared to a fair allocation since VM2 expresses a reduced QoS requirement through its resource bids. Similarly, VM2 experiences an improved allocation when the QoS requirements are reversed. An interesting observation from the figure is that the total allocation r_{cap} is greater in the first half of the experiment compared to the second half. This results from the fact that VM2 consumes more power than VM1 for the same allocation. The system is able to account for this difference through its use of feedback control and reduces r_{cap} based on the current workload mix and allocations.

Finally, Figure 7.11 illustrates the ability of utilizing a richer paravirtualized interface for differentiated service under power budgets. In the experiment, we have four VMs with varying levels of QoS requirements: low, medium, and high. As we move from an unenforced power budget down through high, medium, and low limits, we observe that the application

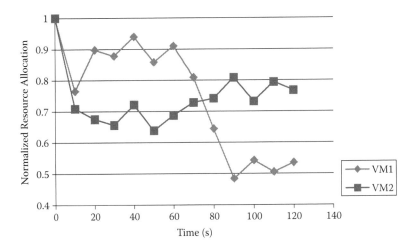

FIGURE 7.10 Trading off resources with varying QoS requirements over time.

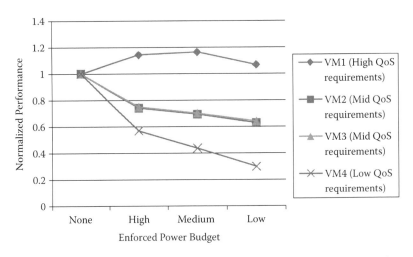

FIGURE 7.11 Support for differentiated services using paravirtualized QoS feedback.

performance experienced by the VMs correlates to the tier of QoS they require. An interesting result in the figure is that the high-priority VM1 actually experiences improved performance as budgets are introduced. This can be attributed to performance interference effects between VMs [26,27], where there is a performance penalty of running consolidated with other VMs on shared multicore resources. As other VMs are capped, these effects diminish improving performance. Overall, these results illustrate the types of management that can be enabled when virtualization infrastructures are extended to incorprate richer paravirtualized interfaces for power management.

7.4 POWER MANAGEMENT MECHANISMS FOR DISTRIBUTED VIRTUALIZED PLATFORMS

In Section 7.3, we discussed the benefits of extending virtualization technologies for improved platform-level management. In large-scale systems, however, management goals are often based on metrics that encapsulate multiple servers. Hence, achieving these goals requires coordinating across separate virtualized platforms. In this section, we take the specific context of power budgeting to illustrate the need for coordinating hypervisors running across machines based on our VPM tokens system [28].

7.4.1 System Managers for Distributed Power Budgeting

When considering power capping across an aggregation of distributed platforms, there are two global metrics of interest. First, of course, is the power consumption of the servers to ensure that the power cap is observed. The second is a notion of utility achieved across a set of managed nodes. An important point here is that the facility-level utility metric is distinct from the application specific QoS discussed in Section 7.3. The former is concerned with data-center-level trade-offs with regard to application performance (e.g., low-priority versus high-priority customers in a cloud scenario), while the latter is at the level of individual applications. The data center utility may be a function of application QoS, however. For example, the utility of a commercial cloud data center may be based on the revenue achieved, which is a function of meeting the QoS expectations of the hosted customer applications.

Efficiently managing data center power budgets for native systems entails power allocation based on applications assigned to physical servers and dynamic management of power based on actual usage. In particular, power allocations are assigned to physical machines based on the data center utility trade-offs of applications deployed there. Subsequently, the system may reallocate power between machines when certain hosts are not using their power resources and others are constrained. This may even include some compensation mechanisms allowing servers that underutilize power during some period to oversubscribe later. The fundamental difference when a data center becomes virtualized is that each of these management actions must be performed at the level of a VM instead of a physical machine. For example, power allocations to a physical machine are now a function of the data center utility trade-offs of the hosted VMs, which vary dynamically as workloads are migrated between physical machines. Hence, the additional level of indirection created by virtualization places a requirement of additional system managers to manage across physical and virtual abstractions. The VPM tokens solution introduces three components for cluster management: a budget manager, platform managers, and a utility manager.

The budget manager is a centralized entity that is responsible for imposing a power budget on underlying virtualized servers. It decides how to translate a specified power constraint into a set of allocations based on the utility of VMs running on physical servers at a given time. Moreover, it must also account for hardware heterogeneity [20], which may have an impact on the amount of resources a given platform can make available as a function of a local cap. To actually enforce a cap, the budget manager communicates with individual platform managers, which provide the base-level support for power management. Specifically, the budget manager conveys to each platform its relative performance capacity limit. These values are determined using a *proportional allocation* approach; the budget manager allocates capacity to physical servers in proportion to the utility impact of the VMs that they host.

In our system, the platform manager is comprised of the VirtualPower platform-level components described in Section 7.3, with extensions. Previously, a VirtualPower-enabled platform managed power based on VPM events triggered by guests. Now, the platform manager must also consider the input provided by the budget manager. It accomplishes this by taking the capacity allocated to it by the budget manager and mapping that to local component capacity limits based on the use of VPM mechanisms to maximize efficiency (e.g., hard and soft scaling). These resource pools are then allocated to VMs based on their relative utilities, again using a proportional scheme. At any given time, however, a VM may use less than the determined allocation either due to low load or because VPM events dictate that the platform manager may reduce allocations if necessary. In this case, the platform manager must also track unused capacity of a VM so that it may be compensated later. The compensation information correlated with a VM is carried with it as it migrates around a data center.

A final manager in our system is the utility manager. Given the dynamic nature of virtualized infrastructures, it is necessary to disassociate utility information from individual platforms since the set of VMs hosted there changes frequently. Hence, both budget and platform managers need to be able to look up the utility information associated with VMs at any given time. The utility manager serves this role. Having provided a brief overview of the additional management components we extend to manage across multiple virtualized servers, we next describe the unifying mechanism that ties them together, VPM tokens.

7.4.2   Coordinating Data Center Management with VPM Tokens
7.4.2.1   Types of VPM Tokens
VPM tokens act as a means of coordination and currency, similar to the use of ticket currencies for scheduling [29]. To go along with the set of system managers, our system incorporates three types of tokens: VM tokens, compensation tokens, and budget tokens. We next describe each of these in turn.

VM tokens are used by the utility manager to convey trade-offs between VMs hosted in the data center. The tokens are sent to platform and budget managers when they query for utility information. Since both of these managers utilize proportional allocation schemes, the absolute values of these tokens do not matter, only their relative weights when compared across VMs. Hence, when assigning VM token values, the utility manager observes a simple constraint that the sum of all outstanding assigned tokens (e.g., not including tokens that were issued for VMs that have since been turned off) is equal to some constant.

Compensation tokens are used to store and convey information regarding underutilized allocations over time that can subsequently be reclaimed when there is spare capacity. In general, compensation tokens can be associated with either physical platforms or VMs, allowing system power management policies to utilize compensation heuristics at physical and virtual levels. In the context of compensation tokens for VMs, compensation tokens can be converted like currency when VMs are migrated across heterogeneous platforms. For example, underutilized power accumulated by a VM on a power-hungry server can be converted into a proportionally larger resource capacity if the VM is subsequently migrated to more efficient hardware. This allows the system to award VMs for "power-friendly" behavior over time.

Our third and final token is the budget token. Budget tokens are used by the budget manager to convey constraint information to distributed platform managers. One of the benefits of abstracting information in this manner as opposed to having the budget manager directly specify constraints in terms of power is that it can help alleviate the budget manager from dealing directly with the complexities of heterogeneous hardware. In particular, instead of using raw power values for budget tokens, the values used are normalized numbers that convey the performance capacity to which a platform must limit itself. However, because of this, we require token conversion models that can be used by both the budget and platform managers. The benefit, though, is that the budget proportional allocation algorithm can deal with normalized budget tokens and disassociate itself from the identification of the conversion trade-offs. In summary, we provide an overview of the VPM token management architecture in Figure 7.12. Next, we describe aspects of how budget tokens are calculated and used by the system.

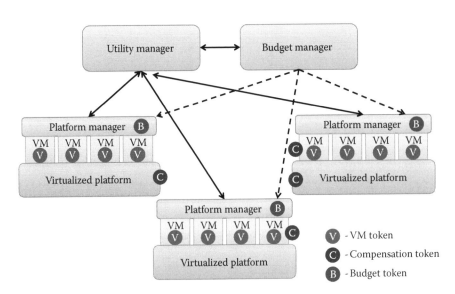

FIGURE 7.12 Distributed power-budgeting architecture with VPM tokens.

7.4.2.2 Managing Power with Budget Tokens

To support conversion of budget tokens, we define for each platform a token-to-power conversion function $F_i(T)$ that gives a platform's power consumption value as a function of a token value T_i. The performance capacity allowable for a platform scales with the token value, and $F_i(T)$ provides the related power value. Based on this definition, the relationship between a budget being imposed by a manager and the set of N token values for platforms can be described by Equation 7.3. In general, $F_i(T)$ can be determined based on off-line profiling of hardware or online tuning.

$$P_{Budget} \geq \sum_{i=0}^{N-1} F_i(T_i) \tag{7.3}$$

As mentioned previously, the budget manager communicates with the platform and utility managers to calculate and distribute appropriate budget tokens to the underlying virtualzed platforms. The budget manager queries all underlying platform managers for the set of VMs that they are running and correlates VM token information from them using the utility manager. These values, along with knowledge of where VMs are running, allow the budget manager to calculate a per platform aggregate utility value U_i. The budget manager then uses the values to weight the tokens assigned to platforms using the token conversion function. Finally, each platform manager uses the platform capacity dictated by the budget token it is sent to perform weighted resource allocations under the budget, in proportion to the VM tokens associated with hosted guests. Both managers may also assign unused power capacity as dictated by the relevant compensation tokens. Based on the overview provided, we next present representative experimental results from a VPM token-enabled system.

7.4.3 Experimental Results

We evaluated VPM tokens using a rich experimental setup consisting of heterogeneous platforms with varying power consumption characteristics. We deployed a mix of virtualized workloads onto the system, including batch workloads, transactional applications, and web service applications. Each of the workload types was assigned VM tokens with different values. We then exercised the budget manager across a variety of budgets in Figure 7.13. The figure compares the overall normalized utility achieved under the power budget with VPM tokens when compared to a default

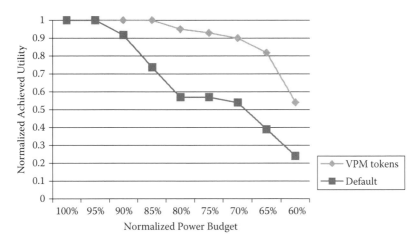

FIGURE 7.13 Data center utility across power budgets with VPM tokens.

policy that uniformly requests platforms to reduce power consumption by the amount required to achieve the budget.

The data in the figure clearly demonstrate the benefits of coordinating power management through the virtualization stack using VPM tokens. Including the additional system managers and VPM tokens allows the system to limit utility degradation to within 10% when the budget is reduced as much as 30%. Moreover, when comparing to the default case, VPM tokens improved utility by up to 43%. These results emphasize the benefits that can be achieved when managing power across distributed hypervisors using the system managers along with budget and VM tokens. To demonstrate the types of benefits that can be enabled with compensation tokens, we take a closer look at the power consumption of a web service application as it is managed over time in Figure 7.14.

Figure 7.14 tracks the power consumed by a web service application as it transitions between periods of load. In the first high-load phase, the application is constrained based on its allocation to meet the cluster power budget. As it transitions to a period of low load, it consumes less than its allocation and thereby accumulates compensation credits. Subsequently, the application moves back into a high-load phase but is still constrained as there is no underutilized capacity in the system. After some time, however, another application underutilizes its allocated power, allowing the web service to consume more than what it would normally be allowed to based on its compensation token. These results further motivate the benefits

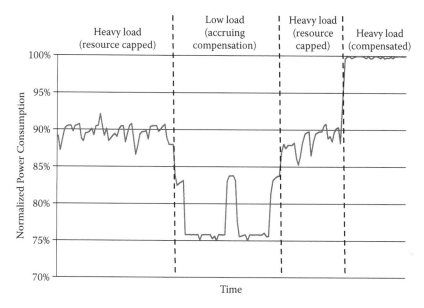

FIGURE 7.14 Compensation with varying load during power budgeting.

of extending virtualization infrastructures for managing power across distributed virtualized servers.

7.5 RELATED WORK

The growing deployment and importance of virtualization has placed significant emphasis on resource management of virtualized servers across all dimensions, including memory efficiency, workload performance, and power consumption. Intelligent memory management is critical for improving hardware utilization via consolidated VMs and can therefore dramatically impact data-center-level power efficiency. Many of the underlying memory management approaches used in commercial technologies, including content-based page sharing, were developed in the context of the VMware ESX Server [30]. Methods to more aggressively pursue page-sharing opportunities, including at subpage granularities, have since been proposed [31]. Other alternatives include removing the need for online scanning of pages to find identical content by integrating appropriate enlightenments [32]. Extending beyond the platform, VM placement techniques designed to enhance page-sharing opportunities have shown promise as well [33]. Beyond page sharing, extensions for memory balancing have been shown to be critical for virtualized systems [34, 35]. These examples illustrate the strides that have been taken to extend virtualization technologies to be more amenable for memory management. The approaches outlined in this chapter, along with other contributions, including VM energy metering support [36], take a similar strategy of integrating improved power management support for virtualization.

Virtualization provides a level of indirection between applications and physical hardware, enabling flexible and efficient workload management in data centers. For example, live migration technologies [5] introduce the ability to adaptively deploy VMs across distributed physical platforms. This underlying mechanism can be used to consolidate VMs in a manner that minimizes unused resources, possibly using forecasting techniques [37,38]. Approaches that take into account the overheads of migrations and plan appropriate sequences to achieve a desired allocation have been developed to better exploit migration as well [39]. An interesting artifact of cohosting VMs is that it creates opportunities for performance interference between VMs due to contention on shared resources [40, 41]. Placement techniques can take this into account and, for example, place workloads to avoid cache contention [42]. A more general framework that overprovisions VMs based on performance interference in cloud environments has been proposed as well [27]. All of these techniques make use of virtualization to better manage data center resources and have an impact: the evolution of system virtualization technologies. Similarly, this chapter demonstrates how power can be managed as a first-class resource in data centers and advocates for continued refinements to virtualization software for energy efficiency.

Managing large-scale systems using control theoretic techniques is a promising direction of research. The use of control theoretic elements for memory balancing on virtualized servers has been explored [34]. Others have incorporated feedback control for resource allocation for multitier applications [43], including management across multiple resource dimensions [44]. The control theoretic techniques presented in this chapter involve just one of many approaches being explored for applying feedback for managing power and performance in data centers. Other work has considered the more general problem of coordinating multiple, possibly independently designed, feedback controllers to jointly optimize both power and performance [45, 46]. As the integration of these techniques becomes commonplace, virtualization extensions like those explored in this chapter can help coordinate across the distributed controllers that will exist in the system [47].

7.6 CONCLUSIONS AND FUTURE CHALLENGES

This chapter has shown ways in which modern virtualization infrastructures can be extended to permit effective power management for large-scale data center systems. The first key insight presented in the chapter is the need for and utility of new interfaces between the virtualization layer controlling physical hardware and the operating systems running applications in guest VMs. The purpose of such interfaces is twofold: (1) to inform hypervisors of the power states acceptable to operating systems and thus to the applications they run and (2) to provide implicit or explicit feedback to operating systems about the efficacy of their internal management strategies. Interestingly, both implicit and explicit feedback can be used, implying that effective power management is possible for guest operating systems whether or not they are aware of running in virtualized environments. However, advantages in terms of power efficiency exist for aware versus unaware systems.

The second insight is that there is a need for new virtualization-level management abstractions. In terms of power management, support is needed for multiple tasks. First, there must be abstractions for exchanging information about and coordinating management actions across multiple data center machines, subsystems, or partitions (sometimes also called "zones") because current hardware is constructed to take per platform and per zone management actions. Without coordination, it is difficult to attain global goals like the enforcement of power caps on zones or entire data centers, and individual decisions can lead to undesirable actions, such as multiple machines being run in lower-power states rather than entirely idling one machine and fully utilizing the other. This is because typically idling machines or turning them off saves more power than running them at lower-power states.

A third insight from this chapter is that virtualization in fact presents new opportunities for power management, not only in terms of the ability to coordinate power management across multiple machines but also because the virtual power states presented to systems and applications need not be identical to actual power states offered by hardware (which tend to be limited). Instead, the virtualization layer can use its own innovative mechanisms to create appropriate power states for application use, including those that hide hardware heterogeneity from systems. The last is particularly important in large-scale data centers, where multiple generations of computing hardware coexist in an ever-present rhythm of machine acquisition, deployment, use, and removal.

A question of note is the role and importance of virtualization in high-performance computing, in fact, answering this question gives rise to several important new directions in research on effective system power management. Given impending government mandates or incentives to operate with high levels of power efficiency, even facilities that have not yet experienced limitations on power delivery (e.g., those running the U.S. Department of Energy's petascale machines) must consider new measures to achieve power efficiency in operating their new machines. Further, there already exist many facilities, in both governmental and commercial endeavors, where limited power supplies threaten or prevent needed equipment upgrades (e.g., due to increased densities). In addition, with new exascale machines being planned, dealing with limitations in power delivery has become a key driver of their design, both at a platform level (resulting in the use of heterogeneous processors, including accelerators, to attain acceptable levels of power/performance) and at the system level, resulting in explicit consideration of "entire platform" power consumption, including memory and the interconnect. Finally, in such settings, virtualization is not only going to be present, in terms of the hardware support needed for it to operate efficiently (e.g., Intel's VT instruction set), but also necessary to facilitate shared facility use by the many groups of researchers requiring machine access.

Interesting challenges for future work in power management for virtualized systems are derived from the increasingly heterogeneous nature of individual platforms, already ubiquitously present in the mobile space and now becoming increasingly important for both server and high-performance systems. Questions arise regarding how to manage platform power and idle states while still meeting application requirements, how to manage VM migration when power/performance characteristics of different cores on a single platform and on different platforms can vary radically, and how to coordinate across entirely different data center subsystems, like those used as external interfaces versus for computational tasks versus those used for I/O or large-scale data analysis, and others.

Other challenges for power management are due to the increasing fungibility of facilities and applications embedded in VMs. Cloud computing has the potential to substantially increase the degree of dynamics experienced in data center systems, so that we must also consider the energy used in dealing with these dynamics (e.g., energy consumed by VM migration) versus just considering steady-state application power consumption. In large-scale clouds, operators will not be aware of which applications run on which machines at what times, so that it becomes hard or impossible to manage machines to continue to meet application goals. As a result, needed are methods for run-time application recognition and identification without compromising applications' privacy concerns and without requiring applications to undergo onerous modification before they can be run in clouds.

REFERENCES

1. Amazon Elastic Compute Cloud. http://aws.amazon.com/ec2.
2. Microsoft. Microsoft Azure Services Platform. http://www.microsoft. com/azure.
3. D. G. Murray, G. Milos, and S. Hand. Improving xen security through disaggregation. In *Proceedings of the International Conference on Virtual Execution Environments (VEE)*, 2008.
4. D. Meisner, B. T. Gold, and T. F. Wenisch. Powernap: Eliminating server idle power. In *Proceedings of the International Conference on Architectural Support for Programming Languages and Operating Systems (ASPLOS)*, March 2009.
5. C. Clark, K. Fraser, S. Hand, J. G. Hansen, E. Jul, C. Limpach, I. Pratt, and A. Warfield. Live migration of virtual machines. In *Proceedings of the 2nd ACM/USENIX Symposium on Networked Systems Design and Implementation (NSDI)*, May 2005.
6. P. Barham, B. Dragovic, K. Fraser, S. Hand, T. Harris, A. Ho, R. Neugebauer, I. Pratt, and A. Warfield. Xen and the art of virtualization. In *Proceedings of the ACM Symposium on Operating Systems Principles (SOSP)*, 2003.
7. VMware. VMware ESX. http://www.vmware.com/products/esx.
8. Microsoft. Windows Server 2008 R2 Hyper-V. http://www.microsoft. com/hyperv.
9. Microsoft. Windows Virtual PC. http://www.microsoft.com/windows/virtual-pc.
10. F. Bellard. Qemu, a fast and portable dynamic translator. In *Proceedings of the USENIX Annual Technical Conference*, 2005.
11. X. Fan, W.-D. Weber, and L. Barroso. Power provisioning for a warehouse-sized computer. In *Proceedings of the International Symposium on Computer Architecture (ISCA)*, June 2007.
12. S. Govindan, J. Choi, B. Urgaonkar, A. Sivasubramaniam, and A. Baldini. Statistical profiling-based techniques for effective provisioning of power infrastructure in consolidated data centers. In *Proceedings of the EuroSys Conference*, 2009.
13. P. Pillai and K. Shin. Real-time dynamic voltage scaling for low-power embedded operating systems. In *Proceedings of the 18th ACM Symposium on Operating Systems Principles (SOSP)*, October 2001.
14. C. Poellabauer, L. Singleton, and K. Schwan. Feedback-based dynamic frequency scaling for memory-bound real-time applications. In *Proceedings of*

the 11th Real-Time and Embedded Technology and Applications Symposium (RTAS), March 2005.

15. P. Ranganathan, P. Leech, D. Irwin, and J. Chase. Ensemble-level power management for dense blade servers. In *Proceedings of the International Symposium on Computer Architecture (ISCA)*, 2006.

16. C. Lefurgy, X. Wang, and M. Ware. Server-level power control. In *Proceedings of the IEEE International Conference on Autonomic Computing (ICAC)*, June 2007.

17. R. Nathuji and K. Schwan. Virtualpower: Coordinated power management in virtualized enterprise systems. In *Proceedings of the 21st ACM Symposium on Operating Systems Principles (SOSP)*, October 2007.

18. Ripal Nathuji, Paul England, Parag Sharma, and Abhishek Singh. Feedback driven QoS-aware power budgeting for virtualized servers. In *Proceedings of the Workshop on Feedback Control Implementation and Design in Computing Systems and Networks (FeBID)*, April 2009.

19. Hewlett-Packard, Intel, Microsoft, Phoenix, and Toshiba. Advanced configuration and power interface specification. http://www.acpi.info, September 2004.

20. R. Nathuji, C. Isci, and E. Gorbatov. Exploiting platform heterogeneity for power effcient data centers. In *Proceedings of the IEEE International Conference on Autonomic Computing (ICAC)*, June 2007.

21. Lucene. Nutch. http://lucene.apache.org/nutch.

22. J. Moore, J. Chase, P. Ranganathan, and R. Sharma. Making scheduling cool: Temperature-aware workload placement in data centers. In *Proceedings of the USENIX Annual Technical Conference*, June 2005.

23. P. Key, D. McAuley, P. Barham, and K. Laevens. *Congestion pricing for congestion avoidance*. Technical Report MSR-TR-99-15, Microsoft Research, February 1999.

24. R. Neugebauer and D. McAuley. Congestion prices as feedback signals: An approach to QoS management. In *Proceedings of the 9th ACM SIGOPS European Workshop*, September 2000.

25. R. Neugebauer and D. McAuley. Energy is just another resource: Energy accounting and energy pricing in the nemesis OS. In *Proceedings of the 8th IEEE Workshop on Hot Topics in Operating Systems (HotOS)*, May 2001.

26. Younggyun Koh, Rob Knauerhase, Paul Brett, Mic Bowman, Zhihua Wen, and Calton Pu. An analysis of performance interference effects in virtual environments. In *IEEE International Symposium on Performance Analysis of Systems and Software (ISPASS)*, April 2007.

27. R. Nathuji, A. Kansal, and A. Ghaffarkhah. Q-clouds: Managing performance interference effects for QoS-aware clouds. In *Proceedings of the EuroSys Conference*, 2010.

28. R. Nathuji and K. Schwan. VPM tokens: Virtual machine-aware power budgeting in datacenters. In *Proceedings of the ACM/IEEE International Symposium on High Performance Distributed Computing (HPDC)*, June 2008.

29. C. Waldspurger and W. Weihl. Lottery scheduling: Flexible proportional-share resource mangement. In *Proceedings of the First Symposium on Operating System Design and Implementation (OSDI)*, 1994.

30. C. Waldspurger. Memory resource management in VMware ESX server. In *Proceedings of the Symposium on Operating Systems Design and Implementation (OSDI)*, December 2002.

31. D. Gupta, S. Lee, M. Vrable, S. Savage, A. Snoeren, G. Varghese, G. Voelker, and A. Vahdat. Difference engine: Harnessing memory redundancy in virtual machines. In *Proceedings of the Symposium on Operating Systems Design and Implementation (OSDI)*, December 2008.

32. G. Milos, D. Murray, S. Hand, and M. Fetterman. Satori: Enlightened page sharing. In *Proceedings of the USENIX Annual Technical Conference*, June 2009.

33. T. Wood, G. Tarasuk-Levin, P. Shenoy, P. Desnoyers, E. Cecchet, and M. Corner. Memory buddies: Exploiting page sharing for smart colocation in virtualized data centers. In *Proceedings of the International Conference on Virtual Execution Environments (VEE)*, March 2009.

34. J. Heo, X. Zhu, P. Padala, and Z. Wang. Memory overbooking and dynamic control of xen virtual machines in consolidated environments. In *Proceedings of the IFIP/IEEE Symposium on Integrated Management (IM) Mini-conference*, June 2009.

35. W. Zhao and Z. Wang. Dynamic memory balancing for virtual machines. In *Proceedings of the International Conference on Virtual Execution Environments (VEE)*, March 2009.

36. J. Stoess, C. Lang, and F. Bellosa. Energy management for hypervisorbased virtual machines. In *Proceedings of the USENIX Annual Technical Conference*, June 2007.

37. N. Bobroff, A. Kochut, and K. Beaty. Dynamic placement of virtual machines for managing SLA violations. In *Proceedings of the 10th IFIP/IEEE International Symposium on Integrated Network Management (IM)*, 2007.

38. G. Khanna, K. Beaty, G. Kar, and A. Kochut. Application performance management in virtualized server environments. In *Proceedings of the 10th IEEE/IFIP Network Operations and Management Symposium (NOMS)*, 2006.

39. F. Hermenier, X. Lorca, J.-M. Menaud, G. Muller, and J. Lawall. Entropy: A consolidation manager for clusters. In P*roceedings of the International Conference on Virtual Execution Environments (VEE)*, 2009.

40. R. Iyer, R. Illikkal, O. Tickoo, L. Zhao, P. Apparao, and D. Newell. Vm3: Measuring, modeling and managing VM shared resources. *Computer Networks*, 53(17): 2873–2887, 2009.

41. Younggyun Koh, Rob Knauerhase, Paul Brett, Mic Bowman, Zhihua Wen, and Calton Pu. An analysis of performance interference effects in virtual environments. In *IEEE International Symposium on Performance Analysis of Systems and Software (ISPASS)*, April 2007.

42. Akshat Verma, Puneet Ahuja, and Anindya Neogi. Power-aware dynamic placement of hpc applications. In *Proceedings of the International Conference on Supercomputing (ICS)*, 2008.

43. P. Padala, K. Shin, X. Zhu, M. Uysal, Z. Wang, S. Singhal, A. Merchant, and K. Salem. Adaptive control of virtualized resources in utility computing environments. In *Proceedings of the EuroSys Conference*, 2007.

44. P. Padala, K.-Y. Hou, K. Shin, X. Zhu, M. Uysal, Z. Wang, S. Singhal, and A. Merchant. Automated control of multiple virtualized resources. In *Proceedings of the EuroSys Conference*, 2009.

45. J. Kephart, H. Chan, R. Das, D. Levine, G. Tesauro, F. Rawson, and C. Lefurgy. Coordinating multiple autonomic managers to achieve specified power-performance tradeoffs. In *Proceedings of the IEEE International Conference on Autonomic Computing (ICAC)*, June 2007.

46. Xiaorui Wang and Yefu Wang. Co-con: Coordinated control of power and application performance for virtualized server clusters. In *Proceedings of the 17th IEEE International Workshop on Quality of Service (IWQoS)*, Charleston, SC, July 2009.

47. R. Raghavendra, P. Ranganathan, V. Talwar, Z. Wang, and X. Zhu. No "power struggles: Coordinated multi-level power management for the data center. In *Proceedings of the International Conference on Architectural Support for Programming Languages and Operating Systems (ASPLOS)*, March 2008.

CHAPTER 8

Demand Response for Computing Centers

Jeffrey S. Chase

CONTENTS

8.1 INTRODUCTION

The term *demand response (DR)* refers to policies or procedures to influence the timing or location of power demand in response to signals from the electricity supplier about energy production cost or availability. DR is an important element of "smart grid" initiatives to improve the reliability and efficiency of electrical power grids.

DR is a form of *demand-side management*, a term that refers to any means to manage the balance of electricity supply and demand in an electrical grid by influencing or modulating electricity demand instead of or in addition to the conventional approach of modulating supply. DR is complementary to demand-side energy efficiency, another form of demand-side management.

Effective demand-side management can reduce environmental impact and operating cost for energy consumers. For example, energy-efficient computing, the primary focus of this book, influences demand by reducing the amount of energy consumed to perform a computational task. Advances in energy efficiency of computing centers reduce their operating costs and environmental impact in an obvious and direct way: Each unit of energy not consumed is one less unit to generate, transmit, and pay for. In particular, the "negawatts" saved by energy efficiency can substitute directly for megawatts produced by burning dirty and expensive fossil fuels [1].

DR offers similar benefits in an indirect way. In contrast to energy-efficient computing, the purpose of DR is not to reduce the amount of energy consumed for any given computing task. Rather, the purpose of DR is to reduce the cost for each unit of energy consumed by controlling when and where that unit is consumed in order to consume it at a time and place with a low unit cost for energy. DR for computing centers involves scheduling or placement of computing loads in a way that considers the availability and cost of the electricity to run those loads. The cost metric may incorporate electricity prices, environmental impact, or other measures.

The role of DR in "green high-performance computing (HPC)" reflects a holistic view of computing and the electricity supply grid as an end-to-end system. In this holistic view, the ultimate measure of energy efficiency is the value of service delivered per unit of fuel consumed or pollution produced. The value derives from the benefit that the information technology (IT) service provides to its users (IT value). Effective DR can enhance energy efficiency on the supply side, even if it does not reduce the amount of electricity needed to produce a given unit of IT value. In particular, DR strategies can enhance end-to-end efficiency by shifting the electricity demand away from dirty electricity generators and onto clean energy or by using energy opportunistically that might otherwise be wasted. DR strategies are also essential to functioning within supply constraints caused by power budgets [2,3], brownout events [4], or intermittent generation [5,6] (e.g., local solar or wind power). Another form of DR is migrating workload in an Internet-scale service to exploit price disparities in regional electricity markets [7].

One challenge of DR is that it often involves trade-offs in the value of service produced. In general, making computing systems more energy efficient enables them to produce the same IT value with less energy and hence lower operating cost. In contrast, DR strategies entail some measurable reduction in service quality and therefore may reduce IT value. For example, a DR strategy might incorporate admission control—the choice to deny or cancel a request for computing service during a period of high energy cost. A DR strategy might also defer or throttle a task or migrate it to a remote provider; any of these choices could reduce the IT value by increasing response time. Another alternative is to reduce the demand for computing power by degrading result quality [8,9].

Thus, DR planning for computing facilities and data centers requires a careful consideration of the impact on IT value. In general, DR strategies are most suitable for what we might call *delay-tolerant computing*. For example, batch job workloads in HPC environments may be less sensitive to response time than interactive web services or other data center applications.

Several intersecting trends suggest that effective DR will be an important design goal for automated load management in computing centers that draw their electrical power from future smart grids. This chapter addresses the following questions:

- How does DR enhance energy efficiency on the supply side? Section 8.2 summarizes the role of DR in "greening" the electrical system to reduce fossil fuel use and carbon emissions.

- How does DR reduce electricity costs for facilities that can shift loads? Section 8.3 gives an overview of electricity pricing models and trends that increase the incentives for adaptive load control.

- Are computing facilities and data centers promising targets for DR strategies? Section 8.4 gives an overview of some factors and trade-offs that determine their suitability and potential to employ DR.

- What factors influence the potential cost savings from DR in computing facilities? What impact does DR have on service quality? Section 8.5 develops a simple analytical model to understand the trade-offs inherent in DR strategies for batch job scheduling. In particular, it illustrates the key factors that influence DR effectiveness in computing centers:

facility load factor (utilization), surplus capacity, facility-scale energy proportionality, and electricity pricing factors.

- How do other changes to energy practices for computing facilities interact with DR? Section 8.6 discusses the impact of advances in facility-scale energy proportionality and dynamic pricing of cloud computing services.

8.2 DEMAND RESPONSE IN THE EMERGING SMART GRID

DR is motivated by a need to balance electricity supply and demand at all levels of the power grid. Electrical grids have little or no energy storage capacity to use as a buffer, so supply must match demand at any point in time. If generation exceeds demand, then energy is wasted. If generation is insufficient to match demand, then outages may occur.

Electricity demand is highly dynamic. Fortunately, electrical demand over a region is predictable with sufficient accuracy and precision to enable a wide range of options for proactive management, including DR strategies. The installed base of electricity-consuming devices changes relatively slowly, and their usage patterns are generally driven by a few primary factors, such as weather, which can be predicted days or hours in advance.

As demand changes, suppliers must modulate generation to match the demand. DR offers a complementary response option: If demand exceeds supply, then reduce demand from selected electrical devices to match the current supply instead of or in addition to increasing supply to meet the demand. DR offers a potential to improve end-to-end efficiency by avoiding reliance on high-cost generators, which are used primarily during periods of peak electricity demand (Section 8.2.1). DR is also an important tool to manage an electricity supply that is itself increasingly dynamic and difficult to modulate. For example, DR becomes more important as grids incorporate a larger share of fuel-free renewable electricity sources into the generation mix (Section 8.2.2).

8.2.1 Importance of Demand Response for Energy Efficiency

To satisfy dynamic demands, electrical suppliers maintain a mix of generating assets with various properties. As demand increases, suppliers dispatch their generating resources according to a plan that attempts to minimize their overall supply costs. Economic dispatch planning may be influenced by a range of factors, including predictions of how long the demand will last and the cost of transmission from the candidate generating plant to the load.

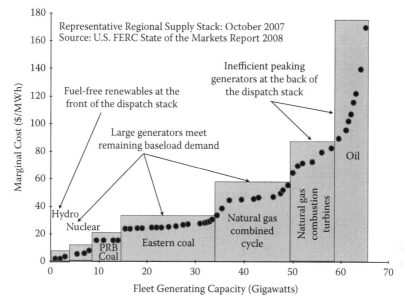

FIGURE 8.1 A representative supply/dispatch stack for a U.S. region with 65 GW of dispatchable generating capacity. The y-axis shows the marginal (e.g., fuel) cost of power from each generator, ordered by their priority in the dispatch stack. The least-efficient generators have the highest marginal cost and are held in reserve for periods of high demand or constrained supply. FERC is the Federal Energy Regulatory Commission. PRB is the Powder River Basin.

Although dispatch planning is complex, the dominating factor is a rank ordering of generators according to least marginal operating cost. The portfolio of generating assets is known as the *dispatch stack*, suggesting a relatively static order of dispatch from preferred plants that run continuously to higher-cost power plants that are used only when needed. Figure 8.1 illustrates a representative dispatch stack [10]. The plants dispatched last are the generators with the highest operating costs for fuel and emissions. These *standby* or *peaking* generators are used only when demands cannot be met from other sources. A 2007 Edison Electrical Institute report suggests that 20% of U.S. generating capacity is used less than 10% of the time [11]. Only 42% of generation capacity is used 100% of the time (this base demand level is known as *baseload*).

Plants designed as peaking plants are intended to be used rarely, so they often lack efficiency features that would increase their capital cost. For example, a typical peaking power plant is a simple gas turbine that is significantly less efficient than combined-cycle gas plants that capture waste heat, as shown in Figure 8.1. The combined-cycle gas plants are cheaper to operate than simple gas turbines, but they are more expensive to build. The back of the dispatch stack also includes some of the dirtiest legacy plants.

DR strategies can improve overall efficiency and reliability by limiting the peak and reducing the use of inefficient standby generation. For that reason, the U.S. Energy Independence and Security Act of 2007 (EISA) mandated comprehensive planning and assessment of DR options for the electrical grid in the United States [12]. The 2009 U.S. National Assessment of Demand Response Potential [13] suggests that DR strategies have potential to enable a 10–20% reduction in peak electricity demand relative to current projections, rendering 188 GW of reserve generating capacity unneeded in 2019. These reductions could allow earlier retirement of legacy assets and freeup resources and capital for investments in clean energy and energy efficiency.

8.2.2 The Role of Renewable Energy

Increasingly, the generating mix is being supplemented with the subclass of "renewable" assets that harvest natural energy flows such as wind and solar, rather than consuming fuel to generate power. Wind plants now make up almost half of newly installed capacity in the United States, and fuel-free renewables are the fastest-growing class of new capacity [10]. They have high capital cost relative to fossil fuel plants, but once installed they, incur no costs for fuel or emissions.

Fuel-free renewables increase the importance of automated DR for two reasons. First, their near-zero operating cost places them at the front of the dispatch stack: By providing clean energy for free, they increase the relative (marginal, unburdened) cost of serving loads with fuel-driven generators. In turn, this effect increases the relative benefit of damping the peak demand. Second, fuel-free power generators are intermittent, and it is not possible to control their output by modulating an input flow of fuel. These properties suggest that the burden of modulating the balance must shift to the demand side as they become more prevalent.

In principle, a computing facility under automated control can modulate its power demand at a fine time granularity to match a dynamic power budget. Researchers have begun to speculate how future DR strategies could play a role in accelerating deployment of renewables collocated with computing centers [5, 6]. These ideas are a first step to developing server backbone infrastructure that can continue to function, perhaps in a degraded mode, if access to fuel-generated power is disrupted.

Another relevant property of fuel-free renewables is that their capital cost is roughly linear with capacity even in small installations; thus, they disrupt the economies of scale that motivated large, centralized generators in the past. Amory Lovins and other leading energy analyists have argued forcefully that this incremental scalability acts against inherent "diseconomies of scale" in centralized electricity generation and distribution [14]. Small-scale deployments distribute capital costs for generating assets, make use of the fragmented available space (e.g., rooftop solar), and reduce transmission costs and losses. They are also the building blocks of "smart microgrids" that can meet local power demands autonomously in the event that the supply of power from the grid backbone is disrupted [15, 16]. To encourage investment in distributed generation, some states have enacted *net metering* laws and *feed-in tariffs* that allow small private renewable energy systems to provide their surplus power to the grid for credit or payment.

These various factors should continue to drive the future power grid toward a larger number of distributed, smaller-scale, weakly controlled, intermittent power sources. In turn, that will add pressure on smart grid control software to balance the increasingly dynamic supply with the dynamic demand. This prospect suggests that DR will become an increasingly important element of integrated control strategies.

8.3 ELECTRICITY PRICING: A VIEW TO THE FUTURE

DR policy choices are driven by conditions in the power network (e.g., congestion, unanticipated demand, changes in supply output, or failure of assets for generating or transmitting electricity). Therefore, a DR strategy requires some stream of information about current or anticipated conditions in the power network. This information acts as a feedback signal from the electricity supplier to the consumer to modulate the consumer's demand.

The nature of the feedback signal is defined by the contract between the electricity supplier and consumer. Some service contracts allow the supplier to modulate demand directly within certain bounds in return for a lower tariff rate (Section 8.3.1). A more flexible feedback signal is a variable electricity price that reflects real-time supply-and-demand conditions (Section 8.3.2). Electricity contracts with hybrid forms of variable pricing are common in the electricity market today, reflecting various balances in the allocation of cost and risk among suppliers and consumers (Section 8.3.3). These contracts continue to evolve.

One premise of this chapter is that computing centers will have increasing exposure to variable pricing for power in the future and will increasingly use DR as a tool to manage their costs and risks. For example, given an adaptive load control algorithm to curtail demand during price spikes, a consumer may lower overall electricity costs by taking on more of the supplier's price risk, in return for a lower average price.

It is also common for electricity contracts to include a charge for the customer's peak demand over a billing period in addition to the energy usage charge. For example, a contract might specify a per kilowatt charge for the average demand over the 15-minute sampling interval with the maximum average demand among all sampling intervals in the billing period. For these contracts, DR strategies can also reduce charges by suppressing the demand peaks.

8.3.1   Dispatchable Demand Response

One simple form of DR contract is an *interruptible tariff*, which grants the supplier (a utility) a right to command the customer to reduce its demand according to prearranged terms. With *direct load control*, the utility issues direct commands to devices on the customer premises (e.g., to modulate systems for heating, cooling, pumping, or battery charging). Alternatively, the customer may simply agree to curtail load to a fixed level or by a fixed amount on command from the provider but retain control over how to meet the target. Customers enter into these agreements in exchange for some payment or pricing incentive [17].

In these agreements, the utility manages the control algorithm to initiate the DR in conjunction with capacity dispatch planning. In essence, the customer's DR commitment is a *dispatchable* resource on an equal footing with generating plants under the supplier's control. In 2008, the U.S. Federal Energy Regulatory Commission issued several regulatory changes to treat dispatchable DR resources comparably to new generating capacity with respect to market function and dispatch planning [10].

Dispatchable DR agreements are most suitable when the DR policy choices made by the utility have negligible impact on the customer. In some electrical devices, demand may be scheduled or shifted in time for short periods without impairing the function of the device. For example, consider a device that has a target running time over specific time intervals, such as a system for battery charging. A control algorithm can modulate the duty cycle over shorter time intervals without missing the target. Other energy-hungry devices maintain a buffer against a leakage or drain rate; examples include pumping systems to maintain a water reservoir level, thermal control in buildings, water heaters, or refrigeration. For these devices, modulating the duty cycle may cause the system to drift from a target objective, but this drift is acceptable within certain tolerances. These systems can be made more DR tolerant by extending the buffer in some way, for example, by increasing the size of the reservoir or by adding insulation or thermal mass.

In contrast, DR for computing services involves managing service quality trade-offs that may be dependent on the applications or load conditions within the center (Section 8.4). It is more suitable to arrangements that allow the center operator to control these trade-offs. Even so, dispatchable DR arrangements are already present in the data center market. For example, some companies (e.g., EnerNOC, http://www.enernoc.com) act as third-party curtailment service providers to broker dispatchable demand reductions and mediate between data center operators and electrical utilities in managing peak loads.

8.3.2 Variable Pricing

A more general alternative to drive DR strategies is to offer variable pricing that reflects varying supply costs through time to the customer. This approach gives less control to the utility, but it offers more flexibility to the customers to manage their own demand.

Variable pricing is a foundation of smart grid technologies. Wholesale electricity markets with dynamic pricing are currently operating in most regions of the United States. These competitive wholesale markets, administered by regional transmission organizations (RTOs) or independent system operators (ISOs), serve more than two-thirds of U.S. electricity customers [18]. These markets use bidding protocols to set a dynamic price on electricity for delivery over specific time intervals within a given transmission region (e.g., on an hourly basis or for spot intervals as short as 5–15 minutes).

While some very large computing centers may purchase electricity in the wholesale market, effective DR generally requires dynamic pricing in the retail markets where the vast majority of end users obtain their power. Retail pricing is decoupled from wholesale prices in most regional

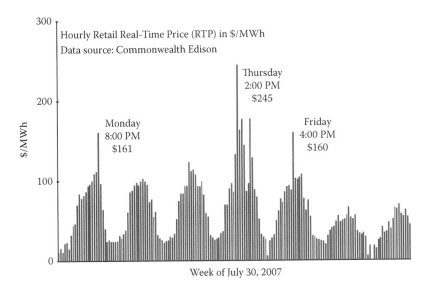

FIGURE 8.2   Real-time prices (RTPs) are affected by diurnal activity cycles and hot weather. Peak demand and peak prices often occur on weekday afternoons. Prices during this hot summer week varied by an order of magnitude.

electricity markets in the United States; in 2009, penetration of dynamic (real-time) pricing at the retail level was still insignificant [12]. This decoupling is largely an artifact of older regulatory regimes that emphasized stable and predictable electricity pricing for consumers. The regulatory climate is changing to integrate more demand-side load management into the grid, including variable-pricing schemes at the retail level [18].

To understand why, consider the effect of fixed-price regimes. Fixed-rate pricing is easy for customers and offers price stability, but providers bear the risk of price swings in the wholesale market. To ensure a profit, they must set the fixed-rate price at a sufficiently high level to balance this risk: The fixed price must be higher than the demand-weighted average of the wholesale price, or the retail supplier loses money. Thus, the retail price must reflect not only the marginal cost of generation but also the risk of supply constraints and price spikes in a dynamic wholesale market.

One straightforward variable-pricing scheme is to pass the wholesale price directly to the consumer, such as by deriving the retail price from the wholesale price according to some preagreed function. This dynamic pricing is known as retail real-time pricing (RTP). Figure 8.2 shows retail prices from an RTP pilot in the state of Illinois: Retail prices fluctuate by the hour according to market conditions, and customers are notified by SMS (Short Message Service) or e-mail before the end of the business day if prices will exceed some user-specified threshold at any time during the following day. Since RTP customers take the risk of price fluctuations in the wholesale market, they should see lower average prices. Although they are exposed to price spikes, they have an opportunity to reduce their costs by limiting their usage during high-price periods. Even at the residential level, price-responsive demand reductions have the potential to damp wholesale price spikes [18], reducing costs for the market as a whole. The number of retail market suppliers offering RTP options to their customers increased by two-thirds between 2006 and 2008 [13].

One limiting factor for RTP and other forms of variable pricing is that they require advanced metering infrastructure (AMI) to monitor customer usage through time. Standard old-style electricity meters measure cumulative consumption but do not record when the consumption occurred. This missing information is needed to bill the customer under a variable-pricing regime. Metering devices that account usage through time had only about 5% penetration in U.S. electricity markets in 2008 [13]. The U.S. government has provided various incentives for AMI deployment, beginning with the Energy Policy Act of 2005 (EPAct).

8.3.3 Hybrid-Pricing Models

Where variable pricing is available, various pricing and contract models have evolved that combine the stability of fixed pricing with the dynamic DR incentives of RTP to varying degrees [11]. Variable-pricing schemes and incentives may incorporate any of several common peak-pricing elements or blend them to distribute costs and risks between the provider and consumer.

- *Time of use* (TOU) is a predictable form of variable pricing with fixed price levels over specific recurring time periods that are designated in advance according to a schedule. The price schedule may be a standard tariff for customers of a given class (e.g., residences) or a negotiated schedule tailored to specific customers and their demand levels. TOU pricing is already common for commercial and industrial (C&I) consumers in many regions of the United States. Figure 8.3 shows the price schedule for a TOU tariff for light commercial customers of Pacific Gas & Electric during summer 2009. Basic TOU pricing reflects only

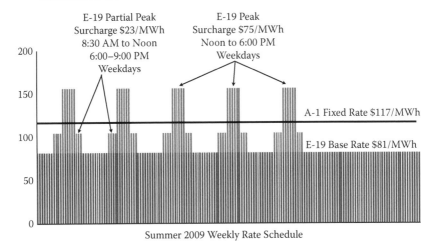

PG&E E-19 Tariff, Time-of-Use (TOU) Pricing
Data source: PG&E Tariff Book

FIGURE 8.3 Electricity may be purchased on a tariff plan that varies prices according to the time of use (TOU) on a regular schedule. TOU pricing enables the customer to plan usage around scheduled surcharge periods that coincide with likely demand peaks.

those wholesale price variations that are anticipated at the time the schedule is set; the supplier bears the risk of any unexpected variation in the wholesale price and must factor this risk into the TOU price levels.

- *Critical peak pricing* (CPP) imposes a surcharge during intervals designated by the provider as "critical" due to unexpected supply constraints. CPP is more dynamic than pure TOU, but the supplier must call critical periods with a minimum advance notice (e.g., a day ahead or an hour ahead), and the contract may limit the number of CPP intervals and the CPP price levels. CPP shifts more of the risk of critical periods to the customer, so it should reduce prices during noncritical periods.

- *Customer baseline load* (CBL) contracts specify a fixed price or schedule for a baseline demand level and different pricing for demand that deviates from the CBL. For example, *block, and index*-pricing is a forward futures contract for a block load at an agreed rate, with demand that deviates from the CBL charged or rebated at the dynamic price.

It is not yet clear how electricity pricing contracts will evolve and what forms they will take in the future. However, there is a clear shift toward more dynamic pricing coupled with incentives for customers that can modulate electricity demand in response to price signals from the electrical grid. The remainder of this chapter assumes that electricity contracts incorporate dynamic pricing for metered usage over specified intervals, and that the consumer controls how and when to modulate its electricity demand to respond to these price signals. To abstract from the pricing alternatives, we may suppose that the customer pays some base rate for electricity together with a surcharge over specific intervals, with both the amount of the surcharge and the surcharge intervals are agreed or announced in advance. It is possible that future models will include competitive bidding for electrical power by large customers, but we do not consider that case further. Pricing factors are discussed in more detail in Section 8.6.2.

8.4 DEMAND RESPONSE AND DEMAND ELASTICITY FOR COMPUTING

Computing centers—supercomputers, data centers, and other server ensembles—offer significant DR potential for the following reasons:

- They are large power consumers, and their share of electrical demand is growing. The analysis by Koomey [19] concluded that their electricity consumption grew at a rate of 16.7% per year worldwide between 2000 and 2005 and by up to 23% per year in some regions. The same paper estimated a growth rate of 12% per year worldwide between 2005 and 2010. A 2007 EPA (Environmental Protection Agency) study [20] projected that the U.S. data center sector would require 5 GW of new peak-generating capacity over the 2007–2011 period under a baseline scenario.

- They have the means to modulate their power demand by controlling the flow of incoming jobs or requests to servers or by suspending, resuming, or migrating work that is already in progress. Servers also have an increasingly rich array of platform-level power actuators under software control, which can select trade-offs of power and performance or cap the power budget at the granularity of individual servers or server ensembles such as chassis or racks [2]. Recent work has shown how to combine and extend these elements to modulate power usage for systems ranging from virtual machines [14] to "warehouse-size" data centers [21].

- They increasingly run automated facility-wide policies to schedule and manage load. These policies can incorporate DR strategies to modulate power demand.

- Networking offers opportunities to shift computing loads and their electrical demand from one region to another. DR-aware load placement can address geographic imbalances of electrical supply and demand, even for interactive services that are sensitive to latency and intolerant of deferring work [7, 22].

The technical challenge for DR in computing is then to extend automated resource management policies to consider electricity cost as an optimization objective. These policies include scheduling, admission control, placement and request routing, and resource control.

Effective DR presumes that demand for electricity by a computing facility is elastic and price responsive. DR strategies respond to higher prices by reducing service, typically substituting service at a later time or a different location. In general, DR for a computing facility compromises service quality by some observable measure. For example, if a DR strategy substitutes off-peak energy use for peak-period energy use, it reduces its demand by deferring work from a peak period to an off-peak period. As a result, any deferred tasks complete later. The degraded service quality is visible through standard measures of responsiveness (e.g., response time or stretch factor), even if the facility has sufficient future surplus capacity to defer work without compromising throughput (see Section 8.5).

A key difficulty is to balance electricity costs against other costs incurred by the candidate response options (e.g., costs to defer, deny, or migrate a computing task). The first challenge is to characterize and predict the impact on service quality. A distinct and perhaps more difficult challenge is to place a monetary value on the degraded service quality so that its cost may be compared directly against the savings in the electric bill.

To make this more concrete, let us assume

1. For a given schedule of activity, the facility consumes electricity over a sequence of discrete time intervals t: $electricity(t)$.

2. The facility incurs a cost for consuming electricity according to a function that varies with time, such as, a base rate plus a variable surcharge: $rate(t)$.

3. For a given schedule of activity, the facility obtains some benefit (IT value) from the work that it completes in each time step. Let us suppose that this benefit can be represented in a common currency to compare it directly with cost: *benefit(t)*.

The DR objective then is to determine a schedule of activity that maximizes the reward:

$$Reward = \sum_t (Benefit(t) - Rate(t) \times Electricity(t)) \qquad (8.1)$$

Consider the common case of a computing facility that serves multiple workload components (e.g., jobs or virtual machines running on behalf of different contending users or groups). For example, cloud hosting centers, enterprise computing centers, and supercomputers execute tasks with a range of priority levels and urgency ranging from mission critical to discretionary. Some workloads offer little opportunity for DR; for example, the IT value of urgent mission-critical tasks is likely to exceed any cost savings of deferring those tasks. Moreover, any new dynamic control incurs some risk of disrupting operations in unexpected ways. As another example, high-throughput computing environments cannot defer valuable work unless they maintain adequate reserve capacity to complete the work later. Section 8.6.1 discusses these practical issues in more detail.

In many cases, such as cloud data centers, the facility is itself a provider that receives revenue from customers according to various service agreements, which may include penalties for violating a service-level objective (SLO). Any scheme for arbitrating resources assigns some relative value to the workloads and uses them to prioritize relative measures to the contending tasks. The difficulty is in mapping these relative measures to an absolute value for the resources they run on and the power they consume. That means quantifying the impact of policy decisions on service quality of each task and the cost of that impact on each component of the workload (e.g., on each customer).

We can think of this challenge in terms of the contract that the facility presents to its customers. The contract may be explicit, as in a service-level agreement (SLA) between a provider and a customer, or it may be implicit in the definition of the service model for the system. In general, the contract imposes some performance constraint or SLO on the facility. For example, the initial contract for Amazon's Elastic Compute Cloud (EC2) suggested that the provider will allocate to each EC2 instance (a virtual machine) all resources that it requests, up to a specified level encoded in the attributes of each instance type. This service model of a minimum *resource entitlement* (or share) is a defining characteristic of *proportional-share* scheduling systems. Alternatively, an SLO may specify constraints on direct measures of application performance, such as bounds on a response time quantile or stretch factor.

If the facility's contract is defined exclusively by such constraints, then the facility is free to allocate any surplus resource as it sees fit once it satisfies the constraints. In particular, the facility is free to allow surplus resources to idle at the discretion of a DR strategy to reduce operating costs. For contracts that specify a penalty for violating the constraints, the DR strategy may choose to violate the constraints and pay the penalty if it is outweighed by other factors [23].

In practice, many computing centers are established by a community to serve its own needs rather than operated for commercial profit with an explicit contract. Today, these systems typically operate on a "best-possible" service model rather than a service constraint. For example, conventional proportional-share service models are defined to be *work conserving*: Any surplus resource is allocated to contending tasks in proportion to their shares, rather than maintained at the discretion of the provider. This means that the user of a proportional-share system has an opportunity to obtain any surplus resources for its own use, competing on a fair footing with other users. Conventional service models with this property are designed with the implicit assumption that the computing resource is a form of public good: Although its use is exclusive, any surplus is free and open for use by the community. For example, the popular Condor job scheduler was orginally conceived as a system to "scavenge" these idle resources [24], which would otherwise be wasted.

DR motivates development of new service models that recognize that the surplus is not free. It is an open question how to design service models that allow the provider to balance the operating cost of surplus resources against the value of using them. In essence, the problem reduces to defining *utility functions* that place value (*benefit*) on service to applications. Several systems have experimented with utility-driven scheduling policies (e.g., [23, 25]), some for the explicit purpose of energy management [4, 26]. It is also intriguing to consider how applications themselves could manage these cost/benefit trade-offs directly through *reflective control*, in which dynamic pricing for cloud service or power is exposed directly to advanced

applications, which respond by modulating their functions and demands
[8,9].

8.5 EVALUATING DEMAND RESPONSE: A SIMPLE MODEL

Consider a system or facility at a single location, executing a workload.
Deferring work during high-cost periods can reduce overall cost to run
the workload, but it incurs a slowdown. Let us consider a simple model to
illustrate the factors that influence the potential for cost savings from DR,
and the resulting slowdown.

This model focuses primarily on a specific example scenario for DR in
computing centers: shifting of batch job workloads in time to minimize
cost under a time-varying electricity price. The example scenario defers
work to take advantage of lower prices in the future, and thus it presumes
that workloads are *delay tolerant* up to some bound. Batch job systems are
an attractive target setting for DR because of their flexibility to schedule
load levels through time given the limited need for interactive response.
However, the principles are relevant to other scenarios as well.

To simplify the analysis, suppose that the cost of electricity varies between two levels, a base price and a peak price, with some given regular
period. Suppose further that the offered workload consists of a continuous stream of arriving jobs that drive the system at a constant load factor.
Figure 8.4 illustrates this scenario. Section 8.6 relates these idealized assumptions to practice.

The model considers a single recurring interval of this schedule, with
parameters normalized to the length of the interval, the base energy price,

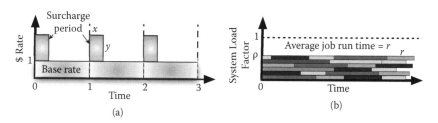

(a)

(b)

FIGURE 8.4 A simple scenario for the analytical model to illustrate demand
response factors. Electrical power is charged at a base rate, with a surcharge
of y times the base rate for critical periods of x time units of each interval
on a regular schedule. The system's offered load is an idealized job mix that
drives the system at a constant load factor ρ with no queuing. The average
job execution time is r.

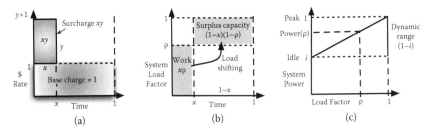

FIGURE 8.5   For the idealized scenario in Figure 8.4, the potential cost savings from demand response is determined by the magnitude y and period x of on-peak electricity surcharges, the system load factor ρ and off-peak surplus capacity, and the system's energy proportionality. We use a linear approximation of power as a function of load: Proportionality is characterized by the *dynamic range* of power consumption from idle i to peak, which is the slope of the line.

and the system's peak power draw, as illustrated in Figure 8.5. Four key factors characterize the potential cost savings and resulting slowdown of DR:

- *Price variability.* Deferring work reduces cost only when it costs less to do the work later. The simplified pricing model consists of a constant *base price* representing a floor on the price of electricity, with a variable *surcharge y*, normalized to the base price, that captures additional costs due to congestion during peak periods, or other factors (see Figure 8.5a). The goal of the DR strategy is to schedule work to avoid these surcharges, subject to various constraints. Higher surcharge rates increase the potential savings from a DR strategy.

- *Surplus capacity.* The system can defer work only if it has spare capacity to run the work later. Without this surplus capacity, deferring work causes monotonically increasing backlogs and slowdowns for later jobs. We characterize the load level of the system as a utilization or *load factor ρ* as a share of its peak capacity to do work: $0 \leq \rho \leq 1$.[1] See Figures 8.4b and 8.5b. DR is an option only when the system is not saturated: average $\rho < 1$.

- *Surcharge time.* A DR strategy defers work from periods of high surcharge to periods of lower (or zero) surcharge. The opportunity for

---

[1]The load factor ρ may be viewed as a measure of IT asset efficiency since it represents the utilization of installed capacity of IT assets [27]. It is analogous to (but distinct from) the *load factor* as the term is used in the electricity sector; it is the ratio of average power (or output of work or electricity) to the peak power (or capacity to do work or generate electricity).

benefit depends in part on the share x of each interval constituting the *surcharge period* during which surcharges apply (see Figures 8.5a and 8.5b).

- *Energy proportionality.* Deferring work can reduce cost only if the system draws less power when it is doing less work. The system power draw is a function of its instantaneous utilization or load factor ρ: $power(\rho)$. Suppose that the system draws a base power i when it is idle, where i is given as a share of the system's peak power. Then, $power(\rho)$ ranges between i and the peak, normalized as 1. The energy proportionality of the system can be characterized by its *dynamic range* $1 - i$ [28]. A dynamic range of 100% ($i = 0$) corresponds to a fully energy-proportional system.

In this model, the total electricity cost to run the system at full power for one recurring interval is $1 + xy$. Consider the case without DR, in which the system runs at a constant load level ρ. It executes work $x\rho$ during each surcharge period. If the system is perfectly energy proportional ($i = 0$), then the base cost for energy during the interval is ρ, and it incurs a surcharge of $xy\rho$, for a total per interval energy cost of $\rho(1 + xy)$.

If the system is not fully energy proportional, then it is necessary to estimate the amount of power the system can save by shifting some or all of its load over the surcharge period. We consider an idealized model of energy proportionality in which power is linear with load factor ρ: $power(\rho) = i + \rho(1 - i)$. The dynamic range is the slope of the line (see Figure 8.5c). For example, if a system consumes 60% of its peak power even while idling in its lowest-power state ($i = 0.6$), then its power varies across 40% ($1 - i$) of its range as ρ ranges from 0 to 1, and the slope of the line is the dynamic range $1 - i = 0.4$. The linear model of energy proportionality was used in early work on energy management for server ensembles [4]; it was also suggested by the recent paper on energy-proportional systems by Barroso and Holzle [28]. This idealized model roughly approximates to the behavior of current-generation servers, but it also applies at facility scale [21,29] (see Section 8.6.3). By this linear model, if the system runs at utilization ρ, then we approximate its power draw as $i + \rho(1 - i)$; thus, the cost to run the system at utilization ρ for one interval is $(i + \rho(1 - i))(1 + xy)$.

Now, consider a DR strategy in this idealized setting. If the DR strategy can defer the $x\rho$ work to a subinterval in which no surcharge applies, then it can idle to consume less power during the surcharge period. During each interval, the system has surplus capacity $(1 - \rho)(1 - x)$ to complete deferred work without incurring a surcharge for the work and without impacting other work scheduled during the interval. To stay idle when surcharges apply, the system must shift $x\rho$ work onto this surplus capacity (refer to Figure 8.5b). It is easy to see that the balance condition reduces to $\rho = 1-x$. If $\rho > 1 - x$, then the DR strategy lacks sufficient capacity to idle during surcharge times: It must run some work even when surcharges apply to avoid creating a backlog. On the other hand, if $\rho \leq 1 - x$, then the DR strategy can idle during the surcharge period. In general, the DR strategy can minimize its costs by shifting $MIN(x\rho, (1 - x)(1 - \rho))$ work and incurs a surcharge for the unshifted residual.

Consider the impact of the DR strategy on energy cost. We can determine an upper bound on the energy cost savings from DR as follows: Suppose that the system has sufficient surplus capacity to shift all of the work $x\rho$ out of the surcharge period, that is, $\rho \leq 1 - x$. If the system is perfectly energy proportional, then it draws zero power while idling ($i = 0$) during the surcharge period, so it can eliminate the surcharge and pay only the base cost ρ for each interval instead of the cost with surcharge of $\rho(1+xy)$. Dividing through by ρ, we have the *idealized savings of DR in an energy-proportional system*, measured as a percentage of energy cost:

$$1 - \frac{1}{1 + xy} \qquad (8.2)$$

In practice, the system is not perfectly energy proportional and consumes some power i even when it is not doing work. This effect reduces the potential savings: A DR strategy can reduce the surcharge incurred but cannot eliminate it. Consider the case where the highest savings occurs: the balance point ($\rho = 1 - x$), where the system idles during each surcharge period and otherwise runs at full power and maximum efficiency. Figure 8.6 depicts this scenario. The system incurs a charge of $ix(1 + y)$ while idling during each surcharge period; it consumes power i at a cost rate of $(1 + y)$ for time x. The off-surcharge cost is again just the base cost ρ: peak power 1 for time $1 - x$ at the base rate 1. Thus, the *best-case idealized savings of DR under the linear power model*, measured as a percentage of energy cost, becomes

$$1 - \frac{\rho + ix(1 + y)}{(i + \rho(1 - i))(1 + xy)} \qquad (8.3)$$

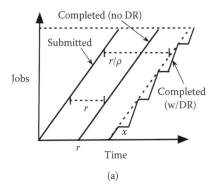
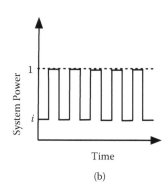

FIGURE 8.6 Job throughput and slowdown (a) and system power (b) under a simple illustrative demand response scenario. Each job completes after exactly r units of running time. The system has just enough surplus capacity to idle during surcharge periods and otherwise runs at full power: $\rho = 1-x$. Average throughput is not affected, but jobs incur an average stretch factor of $1/\rho$.

Figure 8.7 summarizes the interaction of these factors. The figure shows normalized absolute cost savings: how much of the surcharge $xy\rho$ can be avoided. There is no cost to save if $\rho \to 0$ and no opportunity to shift load if $\rho \to 1$. In other cases, the cost and potential savings are linear with ρ: The potential savings grow linearly as the system becomes busier and incurs higher costs but declines linearly when the system has too much load to allow it to idle during the surcharge period. These two lines bound a triangle defining the potential savings. Whatever amount of work is shifted, systems that are more energy proportional (lower i) save more from shifting that work; thus, systems that are not perfectly energy proportional ($i > 0$) obtain savings given by a point in the interior of the triangle rather than on an upper edge. The peak savings for a perfectly energy-proportional system is given by a point on the parabola $yx(1 - x)$: For any given x, the peak savings and top vertex of the triangle occur at the point on the parabola where $\rho = 1 - x$. Thus, the savings of DR are zero if $x \to 0$ (surcharges never apply) or if $x \to 1$ (surcharges always apply). For any point in the triangle, the magnitude of the savings grows linearly with y: Savings are unbounded as y increases.

This cost savings from DR comes at the price of a slowdown as work is deferred to avoid surcharges. The system incurs the maximum average slowdown if it idles whenever surcharges apply: $\rho \leq 1 - x$. For example, consider again the balance point $\rho = 1 - x$ depicted in Figure 8.6. If each

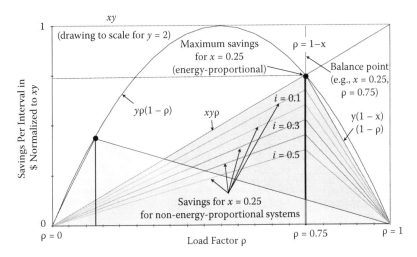

FIGURE 8.7   For any given surcharge time x and surcharge y, the savings is given by a shaded triangle. The system can shift all work out of the surcharge period if $\rho <= 1 - x$: Savings grows linearly with the load factor ρ. At higher load factors $\rho > 1 - x$, the system does not have sufficient spare capacity to idle during the surcharge period: Savings declines linearly with the spare capacity. For any x and ρ, the savings are always proportional to y: Higher y values make the triangle taller. For any given x, y, and ρ, imperfect energy proportionality limits the savings: Higher i values make the triangle shorter. Savings approach zero as $x \to 0$, $x \to 1$, $\rho \to 0$, $\rho \to 1$, or $y \to 0$. The figure is drawn for $x = 0.25$ and $y = 2$.

job requires r units (intervals) of running time to complete, then under the DR strategy it receives $1 - x$ units of service in each interval and requires $r/(1-x) = r/\rho$ intervals to complete. The additional residence time of each job drives the load factor to 1 when the system is active during nonsurcharge periods. The average throughput is unchanged.

This ideal case establishes an upper bound on the slowdown from DR: the stretch factor $1/\rho$. If jobs vary in their run time around a mean of r, then $1/\rho$ is the average stretch factor; some jobs are slowed less, and some are slowed more. In the worst case, a job arrives at the start of a surcharge period and does not quite complete before the next surcharge period: $r = 1 - x$ (plus ϵ). The job completes in time $1 + x$ instead of time $1 - x$, and the worst-case stretch factor for short jobs is

$$\frac{1 + x}{1 - x} \tag{8.4}$$

The worst-case stretch factor grows without bound as $x \to 1$. However, the worst case applies only to the shortest jobs. The maximum run time of a job subject to this worst case is $r = 1 - x$, and $r \to 0$ as $x \to 1$.

8.6 DEMAND RESPONSE IN PRACTICE

The analytical model is useful to illustrate the key factors that influence the effectiveness of a DR strategy. A realistic scenario is likely to be more complicated in several key respects:

- Both the price curves and job properties are more dynamic and often are not known with certainty in advance. For example, a strategy that defers work may expose itself to risk that it will face an unexpected backlog or incur higher costs later.

- The model presumes that the system has the flexibility to suspend, slow, or migrate jobs as needed to implement the strategy, with zero cost. In practice, a DR strategy may have a limited set of actuators, and it must account for their costs.

- The model presumes that the system is unconstrained by the need to manage varying levels of parallelism in jobs. It does not preclude parallel jobs, but it presumes that it can reach any target utilization level by running some subset of its ready jobs. In practice, certain combinations may be infeasible due to the varying resource requirements of jobs.

It is an open problem to develop DR strategies that can manage these factors in practical online scenarios. The benefits of practical DR strategies will approach those derived from the model, although they may be modestly less. It is also important to consider realistic values of the parameters to estimate what these benefits might be in practice.

8.6.1 Load Factor and Capacity Provisioning

The model estimates the cost reductions possible from DR at a given offered load and a given system capacity: The load factor ρ is the ratio of load to capacity. The model shows that DR can reduce costs if $\rho < 1$; that is, the system has surplus capacity.

Recent studies of industry data centers suggested that they have substantial surplus capacity. A recent McKinsey study suggested that many industry data centers with mixed workloads are overprovisioned well beyond their need to handle expected load surges [27]. The study suggested that in many cases structural overprovisioning emerges from organizational factors rather than technical considerations. It argues that a primary goal of data center efficiency efforts should be to reduce capacity to match the load; as a benchmark for these efficiency efforts, it introduces an efficiency metric called CADE (Corporate Average Data Center Efficiency) that is linear with capacity utilization. Overprovisioned systems have high potential for DR, but this benefit has limits; for example, in the model, there can be no advantage to overprovisioning so that ρ is below the balance point $\rho = 1 - x$. Steps taken to improve capacity utilization (higher ρ), such as through server consolidation using virtual machines, do not reduce the potential for DR until they reach this level. Also, there is no energy cost to maintain surplus standby capacity if it is powered off when not in use (see Section 8.6.3).

Another recent study of a well-managed interactive web service showed that servers spent a large majority of their time with CPU utilizations below 50% [28]. However, web data centers have limited opportunity to use their surplus capacity for DR savings by deferring requests. First, most web activity is interactive and so is relatively inelastic: Requests are short, and it would disrupt users to defer them. Second, web service request loads tend to be highly dynamic; studies tend to show regular diurnal request load peaks on weekday afternoons, and flash crowds may also occur. Both electrical grids and web data centers are provisioned with surplus capacity to handle these peaks. Unfortunately, the peaks often coincide: Peak demands on the power grid also tend to occur on weekday afternoons (e.g., when demand is driven by air-conditioning systems). Despite these limitations, recent work has shown that distributed web services have substantial opportunity to reduce electricity costs by routing requests to take arbitrage regional disparities in electricity prices, even given the interactive response constraints [7, 8].

Batch job systems may also have bursty job arrivals, but batch jobs can often be deferred without disrupting users. This makes batch systems more attractive candidates for DR, but it also means they may tend to run at higher load factor ρ. Because response time is less crucial, batch systems have less need for surplus capacity to handle peak loads. These systems tend to be provisioned to sustain the throughput needed to serve a target average load.

In a mission-critical computing center that runs at full utilization, $\rho = 1$, DR offers no cost savings without compromising throughput. However, the center can drive ρ down by investing in surplus capacity. Adding surplus capacity improves average response time to users; with DR, it can also reduce operating costs.

Considering the grid and computing center together as an end-to-end system reveals that investments in computing capacity can be compared directly to investments in peaking generation capacity. For example, suppose a large center runs at full capacity ($\rho = 1$) to serve a given job load, and that the power to run the center is drawn from a grid that experiences a demand spike for 1 h of each day. If the center has 25 racks at 40 kW each, then adding an additional rack permits idling the entire data center during the peak hour without impacting throughput: $\rho = 0.96$. The center delays jobs during the idle period but offers better service for the rest of the day. Idling the center during the demand spike eliminates the need for 1 MW of peaking generation capacity and any fuel that it consumes.

8.6.2 Price Variability

Prices vary within different locations or regions, according to demand, proximity to generating capacity, and the availability and cost of transmission. In 2008, a year of unstable fuel prices, electricity spot prices in the United States fluctuated between \$40/MWh and \$160/MWh. These levels are representative of marginal provider costs (as given in Figure 8.1) but reflect congestion and pricing factors as well [10].

Unusual market conditions occasionally drive real-time market prices well above or below the marginal provider costs; the highest prices exceed the lowest prices by an order of magnitude [11] but may spike above that level in extreme cases. For example, price spikes to \$8,000/MWh have occurred during extreme weather events (northeastern United States in summer 1999). In California in 2000–2001, electricity suppliers drove wholesale prices to regulatory cap levels (\$1,000/MWh). However, it is dangerous to infer too much about price variability from these extreme events in freshly deregulated markets. Indeed, one motivation for DR is that it reduces the market power of suppliers to drive extreme price spikes by withdrawing supply, as apparently occurred in California during the 2000–2001 crisis [30,31].

Figures 8.2 and 8.3 are more likely to be representative of pricing conditions encountered in practice. For the E-19 tariff in Figure 8.3, the DR model parameters are $x = 0.25$ and $y = 0.92$ if we consider only the on-peak periods. The maximum savings is 18%. Considering both on-peak and partial-peak periods, the parameters are $x = 0.54$ and $y = 0.58$; in this case, y is the time-weighted average surcharge for on-peak and partial-peak periods. The maximum savings from DR is 24%, but it can be obtained only if the center runs at less than half of its capacity: $\rho = 1 - x = 0.46$. PG&E's residential A-6 tariff for the same season had a higher on-peak surcharge ($y = 1.55$) and a potential DR savings of 28% for a facility that is loaded at an average 75% of capacity. In the RTP example in Figure 8.2, the top 5% of pricing intervals averages $149/MWh, and the 95% average price is $52. Taking $52 as the base rate, the average normalized surcharge is $y = 1.86$ for $x = 0.05$. A center paying these prices could save 8% even if it is loaded at an average 95% of capacity. Taking the top 2% of pricing intervals as the surcharge period, yields $y = 2.25$, and the maximum savings is 4.3%.

It is important to note that customers can often lower their average prices by accepting the higher risk of volatility that comes with variable pricing. Thus, DR may be viewed as a risk-control measure with an indirect benefit of lowering electricity prices during normal operation while limiting exposure to the resulting price spikes. For example, the aforementioned A-6 tariff offers a discount of 41% on the base price of electricity (11 hours per day), and a 27% average discount on electricity outside peak periods. The customer obtains these benefits by accepting the surcharge for expected peak periods. A DR strategy can avoid the surcharges if it can defer electricity usage during these surcharge periods.

8.6.3   Energy Proportionality at Facility Scale

The model illustrates the importance of energy proportionality for DR savings. In essence, energy proportionality captures the degree to which a system can reduce its power draw by shedding load. The idealized model presumes that the system power is linear with instantaneous facility utilization or load factor ρ. Refer to Figure 8.5c and the discussion in Section 8.5.

We can quantify energy proportionality at the granularity of servers or other individual components or at the granularity of ensembles or an entire facility. For example, recent results from the SPECpower benchmark indicated that server systems are increasingly energy proportional, primarily as a result of advances in CPUs and power supplies. Servers with dynamic ranges of 70% or higher are common. However, their power profiles increasingly deviate from the linear model, which tends to underestimate their power draw at CPU utilization levels that are low but nonzero. Also, for data-intensive workloads, the energy costs of memory, storage, and I/O may dominate CPU activity [32], and these costs tend to be less energy proportional than CPUs.

In server ensembles, further improvements are possible by concentrating load on a minimal subset of servers and stepping down surplus servers to a low-power state (e.g., [4]). This technique can be combined with various approaches to active server scaling at the platform level, such as dynamic voltage scaling. Several commercial products and services offer support for energy-proportional ensembles using these techniques. Recent studies suggested that server ensembles can approach full energy proportionality with active management [21, 29]. Related techniques have been applied in storage ensembles, with some success (e.g., [33]). There has also been some recent attention to energy-proportional networking for data centers [34].

A large share of power in computer centers and data centers feeds ancillary equipment, including cooling and power distribution, rather than servers. One measure of their relative impact is the ratio of total power to power for servers and other IT equipment—the ratio known as power usage effectiveness or PUE. Recent studies have estimated a typical PUE value of 2.0 [19], suggesting that about half of the energy in today's data centers goes to servers. The EPA target for state-of-the-art data centers was a PUE of 1.2 in 2011 [20]. Google reported PUE levels for Google-designed data centers on a quarterly basis and succeeded in meeting a PUE of 1.2 in 2010. Active server management pushes PUE up, making efficient power distribution and cooling more important.

In recent years, energy-proportional cooling has received more attention. For example, temperature-aware workload placement helps reduce cooling demands for ensembles running below full capacity [35]. Other "smart cooling" techniques modulate fan speeds, compressor duty cycles, and other mechanical systems. A recent study suggested that combining these techniques with active server management can yield facility-level energy proportionality roughly following the linear model with dynamic ranges of 70% to 80% [29].

8.7   SUMMARY AND CONCLUSION

REFERENCES

1. Amory Lovins. The negawatt revolution. *Across the Board, the Conference Board Magazine*, 27(9), September 1990.
2. Partha Ranganathan, P. Leech, David Irwin, and Jeffrey Chase. Ensemble-level power management for dense blade servers. In *33rd International Symposium on Computer Architecture (ISCA)*, June 2006.

3. Ripal Nathuji and Karsten Schwan. VirtualPower: Coordinated power management in virtualized enterprise systems. In *Proceedings of the ACM Symposium on Operating Systems Principles (SOSP)*, October 2007.

4. Jeffrey S. Chase, Darrell C. Anderson, Prachi N. Thakar, Amin M. Vahdat, and Ronald P. Doyle. Managing energy and server resources in hosting centers. In *Proceedings of the 18th ACM Symposium on Operating System Principles (SOSP)*, pages 103–116, October 2001.

5. Christopher Stewart and Kai Shen. Some joules are more precious than others: Managing renewable energy in the datacenter. In *Proceedings of the Workshop on Power-Aware Computing and Systems (HotPower)*, October 2009.

6. Navin Sharma, Sean Barker, David Irwin, and Prashant Shenoy. Blink: Supply-side power management in data centers. In *Proceedings of the Sixteenth International Conference on Architectural Support for Programming Languages and Operating Systems (ASPLOS)*, March 2011.

7. Asfandyar Qureshi, Rick Weber, Hari Balakrishnan, John Guttag, and Bruce Maggs. Cutting the electric bill for Internet-scale systems. In *Proceedings of the ACM SIGCOMM Conference*, October 2009.

8. Woongki Baek and Trishul M. Chilimbi. Green: A framework for supporting energy-conscious programming using controlled approximation. In *Proceedings of the 2010 ACM SIGPLAN Conference on Programming Language Design and Implementation*, PLDI '10, pages 198–209, ACM, New York, 2010.

9. Azbayer Demberel, Jeffrey Chase, and Shivnath Babu. Reflective control for an elastic cloud application: An automated experiment workbench. In *Proceedings of the First Workshop on Hot Topics in Cloud Computing (HotCloud)*, June 2009.

10. Federal Energy Regulatory Commission. *State of the Markets Report*, August 2009.

11. Steven Brathwait, Dan Hansen, and Michael O'Sheasy. Retail electricity pricing and rate design in evolving markets, July 2007.

12. Federal Energy Regulatory Commission. *National Action Plan on Demand Response* (Draft), March 2010.

13. David Kathan, Caroline Daly, Jignasa Gadani, Diane Gruenke, Eric Icart, Ryan Irwin, Carey Martinez, Kendra Pace, John Rogers, Christina Switzer, Carol White, and Dean Wight. Assessment of demand response and advanced metering, September 2009.

14. Amory B. Lovins, E. Kyle Datta, Thomas Feiler, Karl R. Rabago, Joel N. Swisher, Andre Lehmann, and Ken Wicker. *Small Is Profitable: The Hidden Economic Benefits of Making Electrical Resources the Right Size*. Rocky Mountain Institute, Boulder, CO, 2002.

15. M. M. He, E. M. Reutzel, Xiaofan Jiang, R. H. Katz, S. R. Sanders, D. E. Culler, and K. Lutz. An architecture for local energy generation, distribution, and sharing. In *Energy 2030 Conference, 2008. ENERGY 2008. IEEE*, pages 1–6, 2008.

16. H. Farhangi. The path of the smart grid. *Power and Energy Magazine, IEEE*, 8(1):18 –28, January 2010.

17. Federal Energy Regulatory Commission Staff Report. *National Assessment of Demand Response Potential*, June 2009.

18. Jon Wellinghoff, David L. Morenoff, James Pederson, and Mary Elizabeth Tighe. Creating regulatory structures for robust demand response participation in organized wholesale electric markets. In *ACEEE Summer Study on Building Efficiency*, August 2008.

19. Jonathan G. Koomey. Worldwide electricity used in data centers. *Evironmental Research Letters*, 3:034008, September 2008.

20. United States Environmental Protection Agency (EPA). *Report to Congress on Server and Data Center Energy Efficiency*, Public Law 109-431, August 2007.

21. Xiaobo Fan, Wolf-Dietrich Weber, and Luiz Andre Barroso. Power provisioning for a warehouse-sized computer. In *Proceedings of the International Symposium on Computer Architecture (ISCA)*, June 2007.

22. Kien Le, Ozlem Bilgir, Ricardo Bianchini, Margaret Martonosi, and Thu D. Nguyen. Managing the cost, energy consumption, and carbon footprint of internet services. In *Proceedings of the ACM SIGMETRICS International Conference on Measurement and Modeling of Computer Systems*, SIGMET-RICS '10, pages 357–358, ACM, New York, 2010.

23. David Irwin, Laura Grit, and Jeff Chase. Balancing risk and reward in a market-based task service. In *Proceedings of the Thirteenth International Symposium on High Performance Distributed Computing (HPDC-13)*, June 2004.

24. M. Litzkow, M. Livny, and M. Mutka. Condor—A hunter of idle workstations. In *Proceedings of the 8th International Conference on Distributed Computing Systems*, pages 104–111, 1988.

25. Alvin Auyoung, Laura Grit, Janet Wiener, and John Wilkes. Service contracts and aggregate utility functions. In *Proceedings of the IEEE Symposium on High Performance Distributed Computing*, pages 119–131, 2006.

26. Michael Cardosa, Madhukar R. Korupolu, and Aameek Singh. Shares and utilities based power consolidation in virtualized server environments. In *Proceedings of the 11th IFIP/IEEE International Conference on Symposium on Integrated Network Management*, June 2009.

27. James M. Kaplan, William Forrest, and Noah Kindler. Revolutionizing data center energy efficiency, July 2008.

28. Luiz Andre Barroso and Urs Holzle. The case for energy-proportional computing. *Computer*, 40:33–37, 2007.

29. Niraj Tolia, Zhikui Wang, Manish Marwah, Cullen Bash, Parthasarathy Ranganathan, and Xiaoyun Zhu. Delivering energy proportionality with non energy-proportional systems—Optimizing the ensemble. In *Proceedings of the Workshop on Power-Aware Computing and Systems (HotPower)*, October 2009.

30. Kathleen Spees and Lester B. Lave. Demand response and electricity market efficiency. *The Electricity Journal*, 20:69–85, April 2007.

31. Severin Borenstein. The trouble with electricity markets: Understanding california's restructuring disaster. *Journal of Economic Perspectives*, 16(1):191–211, 2002.

32. Dimitris Tsirogiannis, Stavros Harizopoulos, and Mehul A. Shah. Analyzing the energy efficiency of a database server. In *Proceedings of the 2010 ACM SIGMOD International Conference on Management of Data*, 231–242, Indianapolis, IN, ACM, New York, 2010.

33. Qingbo Zhu, Zhifeng Chen, Lin Tan, Yuanyuan Zhou, Kimberly Keeton, and John Wilkes. Hibernator: helping disk arrays sleep through the winter. In *Proceedings of the Twentieth ACM Symposium on Operating Systems Principles*, 177–190, Brighton, United Kingdom, ACM, New York, 2005.

34. Brandon Heller, Srini Seetharaman, Priya Mahadevan, Yiannis Yiakoumis, Puneet Sharma, Sujata Banerjee, and Nick McKeown. ElasticTree: saving energy in data center networks. In *Proceedings of the 7th USENIX Conference on Networked System Design and Implementation*, San Jose, CA, 17, USENIX Association, Berkeley, CA, 2010.

35. Justin Moore, Jeff Chase, Parthasarathy Ranganathan, and Ratnesh Sharma. Making scheduling "cool": Temperature-aware workload Placement in data centers. In *Proceedings of the 2005 USENIX Annual Technical Conference*, pages 61–74, April 2005.

CHAPTER 9

Implications of Recent Trends in Performance, Costs, and Energy Use for Servers

Jonathan G. Koomey, Christian Belady, Michael Patterson, Anthony Santos, and Klaus-Dieter Lange

CONTENTS

9.1 INTRODUCTION

As data centers have grown in both economic importance and cost, the need for understanding the underlying drivers of total costs in the data center has also increased. In particular, the relationships among processing power, energy use, and purchase costs of information technology (IT) equipment in these facilities strongly affect the fraction of total costs attributable to IT equipment (as distinct from facilities/infrastructure equipment like chillers and power distribution systems).

Anecdotal reports indicated that infrastructure equipment related to power and cooling may be responsible for about half of total annualized costs in typical data center facilities [1–4], and that this fraction is growing over time as IT equipment acquisition costs decline and IT equipment energy use increases. This finding is surprising to people new to the data center arena, as they associate these facilities mainly with the IT equipment they contain.

Unfortunately, there has been little systematic, transparent, and peer-reviewed work documenting the aggregate trends in IT equipment that are driving changes in total data center costs. This lack is most keenly felt by those trying to plan for new facilities. Modeling data center costs at a high level requires abstracting from anecdotal data to generalize about trends, but the poor quality of available data and examples has prevented such generalizations from being useful to the bulk of the data center industry.

This chapter assesses trends in server equipment (the most important component of IT equipment in data centers) in a way that will be useful for people trying to understand data center costs at a high level. Drawing on a previous analysis [5], it summarizes trends in server costs, energy use, and performance and describes the implications of those trends for the economics of high-density computing facilities.

9.1.1 Conceptualizing the Problem

One of the most important aggregate parameters affecting the cost of data centers is the amount of direct power use (watts) associated with $1,000

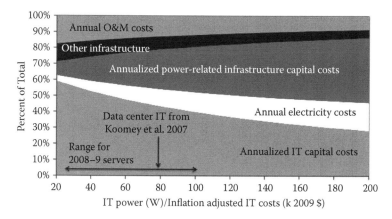

FIGURE 9.1 As power per server costs grow, power-related costs grow in importance. This graph shows annualized costs for a tier 3 data center. The 2008–2009 server data are from Figure 9.2. Capital and operating costs were derived using equations in the Appendix.

of expenditure on IT equipment hardware (in this case, servers[1].) Brill [2] showed (using anecdotal information) that this parameter has been increasing rapidly in recent years, which has made cooling and power infrastructure costs rival the IT capital costs in some recently constructed data centers. If this trend continues, the power-related infrastructure costs will significantly exceed the IT capital costs for new facilities in under a decade, a finding that has implications for how these facilities are built and how their costs are allocated within organizations [1–3].[2]

Figure 9.1 shows how annualized IT capital costs compare to annual electricity and annualized infrastructure capital costs as a function of power use/server cost. The figure uses the equations in the Appendix and the infrastructure cost and electricity price assumptions from Koomey [3] for a tier 3 data center.[3] At 100 W per $1,000 of server cost, IT capital costs

---

[1] In principle, this parameter should be measured for all IT equipment in the data center, not just servers, but the data, sparse as they are, are most available for servers, so that is what we focus on here.

[2] Most companies have separate budgets for IT and facilities expenditures, and if a dollar spent on IT can commit another part of the company to a dollar or more of additional expenditures in a separate budget, suboptimal behavior will generally be the result.

[3] Each data center is unique. They vary greatly depending on the reliability they deliver and the types of computing they support. This example was developed for high-performance computing for financial applications. It is the best-documented published example of data center costs, which is why we relied on it for our discussion here. The conceptual points raised in our discussion are not affected by the specifics of this example, and we believe the concreteness this example lends to the discussion outweighs any potential pitfalls.

represent about 40% of total costs, and at 200 W per $1,000, they are responsible for less than one-third of total costs, which means that for every $1 spent on IT equipment, a company would be committing to at least another $2 for electricity use, power and cooling capital costs, and other costs. These results have implications for assessing and controlling costs in these facilities, as discussed in the future work section that follows.

Power per server cost (in watts per $1,000) can be decomposed into two component parts, as shown in Equation 9.1a:

$$\frac{\text{Power}}{\text{Server cost}} = \frac{\text{Power}}{\text{Performance}} \times \frac{\text{Performance}}{\text{Server cost}} \qquad (9.1a)$$

or equivalently

$$\frac{\text{Power}}{\text{Server cost}} = \frac{\dfrac{\text{Performance}}{\text{Server cost}}}{\dfrac{\text{Performance}}{\text{Power}}} \qquad (9.2b)$$

where

$\dfrac{\text{Performance}}{\text{Power}}$ = System performance divided by the measured power use for that server system to deliver that performance (i.e., performance per watt); and

$\dfrac{\text{Performance}}{\text{Server cost}}$ = That same performance metric divided by the server hardware capital cost as configured to achieve that performance.

This equation explains why measuring power use, performance, and server costs in a consistent fashion is so important: It allows us to understand the underlying drivers of power per server cost in an unambiguous way. It also shows that whenever performance per server cost is increasing faster than performance per watt, power use per $1,000 of server costs will increase.

Consider Figure 9.2, which plots the two components in Equation 9.1b for 14 servers selected from the available SPEC Power runs (Standard Performance Evaluation Corporation 2013) for servers manufactured circa 2009, including the HP ProLiant DL360 G5 machine analyzed by Koomey [5]. For comparison, it shows data for the HP ProLiant DL360 G1 machine from that same report, as well as two lines of constant watts per $1,000 (one for 25 and one for 100).

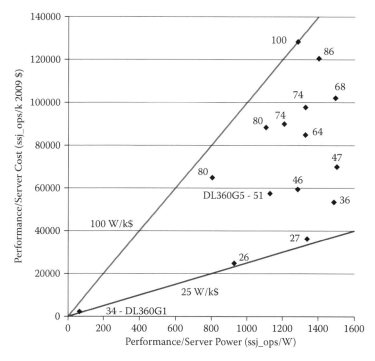

FIGURE 9.2 There are wide ranges of performance per watt and performance per server costs in servers available circa 2009. Performance and power were based on 100% load cases from SPECpower_ssj2008 runs, as documented in Appendix B of Koomey [5]. Numbers next to each data point represent watts per thousand 2009 dollars of IT equipment expenditure for each server. DL360 G1 data (circa 2001) added from Koomey [5] for comparison.

The x-axis plots server performance at maximum load divided by power, and the y-axis plots server performance at maximum load divided by purchase cost. The graph shows almost a factor of two variation in server performance per unit of power and a factor of four variation in server performance per server purchase cost. Combining these parameters yields a range of from 26 to 100 W per $1,000 of server equipment, as shown in Figure 9.1.

These graphs illustrate the complex and multivariate nature of the decision problem for data center design. Increasing performance per purchase cost is a surefire way to reduce the direct cost of delivered computing services, but if performance per watt is low, that choice will exact a penalty in infrastructure capital and electricity costs.

Consider the following stylized cost calculation for the total costs of a data center, based on the simple model developed in Koomey [3] and the equations in the Appendix. The annualized total cost (ATC) of a data center can be expressed as in Equation 9.2:

$$ATC = IT + INF_{kw} + INF_{nonkW} + EC_{IT} + EC_{Inf} + O\&M \qquad (9.2)$$

where

IT = annualized IT capital costs (which include the acquisition costs of servers, network gear, disk arrays, and other IT equipment);

INF_{kw} = annualized kilowatt-related infrastructure capital costs (like those for chillers, water distribution, cooling towers, backup power systems, generators, and anything else whose sizing is dependent on the amount of power drawn by the IT equipment);

INF_{nonkW} = annualized non-kilowatt-related infrastructure capital costs (which include building shell, office fittings, and land);

EC_{IT} = annual direct electricity costs for IT equipment;

EC_{Inf} = annual direct electricity costs for infrastructure equipment (in typical data centers, $EC_{IT} \approx EC_{Inf}$); and

$O\&M$ = annualized operations and maintenance costs (in our definitions, this term includes both IT and facilities operations costs but does not include software licenses and application development).

Both annual electricity cost and kilowatt-related infrastructure costs are directly related to IT expenditures through the ratio of power use per server cost. In the example in Koomey [3], where the aggregate power use per purchase cost for all IT equipment in the data center in 2009 dollars was about 80 W/$1,000, IT capital costs accounted for 45% of total annualized costs, electricity use accounted for about 10%, and power-related infrastructure capital accounted for almost a quarter of the total (see Figure 9.3).

Of course, what we really care about is the cost per delivered computing cycle. Let us think about this problem in terms of the maximum number of computations possible for a given data center over the course of a year.[4]

---

[4]The subtleties of measuring actual utilization and total computational output are complex ones that need not enter into our illustration here. Poorly utilized data centers can of course lower their total costs of computing substantially by increasing utilization levels. Such changes will have a large effect

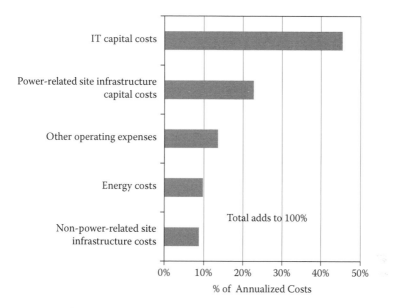

FIGURE 9.3 Infrastructure capital costs and electricity costs are substantial for tier 3 data centers. (Based on [3].)

Dividing both sides of Equation 9.2 by the maximum annual computations, we obtain

$$\frac{ATC}{Annual\ Computations} = \frac{IT + INF_{non\text{-}kW} + INF_{kW} + EC + O\ \&\ M}{Annual\ Computations} \quad (9.3)$$

Equation 9.3 represents in a schematic form the complete decision for data center design. The designer would like to minimize the total cost for delivering computations, but achieving this goal is not as simple as choosing the server with the maximum performance or lowest power use per dollar of equipment purchase cost. Focusing only on the ratio of IT costs per computation would result in a significantly more expensive facility than if the data center were analyzed as a whole system.

9.1.2 Implications of These Equations

These equations, combined with those in the Appendix, can be used to give quantitative insight about the trade-offs among the different cost components of data centers. Let us assume Moore's law drives performance

---

on computational efficiency because current server power use does not generally scale exactly with computational output, and there is a large fixed power draw when the server is idle [6].

per server cost up by a factor of two over a 2-year period (a doubling time of 2 years). The effect on the power-related components of data center costs depends on what happens to power use per server costs (and, implicitly, to performance per watt).

- If performance per watt also doubles during this period, then watts per $1,000 will remain constant (as per Equation 9.1b) and the overall cost per computation will be exactly halved (Equation 9.3) because each of the numerator elements remains constant and the denominator doubles.

- If performance per watt does not change at all, watts per $1,000 will double, and the 50% reduction in cost per computation will become a 32% reduction in total costs (see table in the Appendix for details). In this case, increased indirect power-related costs offset 36% (18%/50%) of the cost reductions resulting from increased compute performance. Another way to say this is that the total cost of building and operating a data center that delivers the increased performance would be 36% higher than in the base case.

- To make the increase in indirect costs exactly offset the benefits from increased server performance, performance per watt would have to drop to almost half of its initial value at the same time as performance per server cost is doubling. This change would result in power use per server cost of more than 300 W/$1,000 in 2009 dollars, almost a factor of four increase over the base case value (about 80 W per $1,000 in 2009 dollars; from [3].

These effects cut in both directions. If server manufacturers were able to triple performance per watt as performance per server cost was doubling, the total cost per computation would be 12% less than if performance per watt just kept pace with performance per server cost because of the reduction in power-related costs. Whether investing to make this change would be economically desirable depends, of course, on the costs to improve server efficiency at this rate.

9.1.3 The Focus of This Study

To understand the underlying drivers for this complex situation, this chapter explores trends in power use, server costs, and performance. It focuses on the following questions:

- What kinds of data would be needed to accurately characterize trends in performance per watt, performance per server cost, and power use per server cost?

- Can changes in these parameters be measured in a credible, accurate, and representative way using publicly available data?

- If so, how have these parameters changed since 2003, and what can we say about how they are likely to change in the next decade?

9.2 DATA AND METHODS

We developed case studies to characterize historical trends in costs, performance, and energy use of servers. One example involved HP's Superdome high-end server; the others focused on what the industry calls "volume servers," including data from Google, Lawrence Berkeley National Laboratory, Intel, and HP. We used consistent performance benchmarks for old and new servers, even going so far as to run the recent SPEC (Standard Performance Evaluation Corporation) power benchmark on some older volume servers. We also used measured data on power use and acquired real data on costs to develop trends. For more details, see Koomey [5].

9.2.1 General Issues

Our purpose here is to report on peer-reviewed consistent comparisons for performance, costs, and energy use over time. By *peer reviewed*, we mean that a broad section of knowledgeable industry observers (identified by name in the Acknowledgments section to this chapter) have examined the assumptions, data, and analysis and found them credible. By *consistent*, we mean that measurements of these parameters are conducted in a fashion that allows for meaningful comparisons over time.

To understand these trends for server equipment, we first need to define system boundaries. Servers can be analyzed at the CPU (central processing unit) level, the system level, or the applications level.[5] The applications level is closest to the tasks that users are performing, but data at that level are the hardest to measure and to generalize. Data are abundant at the CPU level, but CPU measurements are sufficiently removed from actual computing

---

[5]The most sophisticated data center operators that have relatively homogeneous computing loads can analyze servers at the data center level since they can shift loads between servers relatively easily. This system-level analysis is not relevant for most users (who are more concerned with server-level trends), so we do not discuss it further here. It is also important for improving equipment utilization, another topic we do not treat here.

tasks that they are of limited usefulness. System-level data are in the middle in terms of both data availability and relevance to actual computing tasks. In practice, the system-level data are the most likely to be both available and relevant.

It is important to ensure that any examples used be representative of IT equipment. There are at least two dimensions in which server hardware can be representative: configuration and operation. Server systems can be configured with variations in random access memory (RAM), disk drives, and network interface cards—examples chosen should be as representative of typical configurations as possible. Most business servers operate at only 5% to 15% of their maximum computing loads, but there is wide variation in compute utilization. The ideal examples would be broadly representative of the ways servers run actual applications.

To allow straightforward comparisons, we use the metric of doubling time, defined as the number of years it takes for a parameter (performance per watt, for example) to double. We first calculate the instantaneous growth rate g as in Equation 9.4:[6]

$$g = \frac{LN\left(\frac{Y_t}{Y_0}\right)}{t} \tag{9.4}$$

where Y_t is some quantity at time t, Y_o is that quantity at time 0, and t is the time over which growth occurs, measured in this case in years (from year 0 to year t).

Instantaneous growth rates assume continuous compounding, which is necessary when dealing with the rapid growth rates common in computer technology. An instantaneous growth rate of 69.3% implies a doubling every year.

We can then calculate the doubling time using Equation 9.5:

$$Doubling\ time = \frac{LN\,(2)}{g} \tag{9.5}$$

---

[6]It is more common in most situations to use simple growth rates, calculated as $g = \left(\frac{Y_t}{Y_0}\right)^{\left(\frac{1}{t}\right)} - 1$, but this method gives erroneous answers for growth rates higher than about 10% per year. For the high growth rates common to IT equipment, instantaneous growth rates are more appropriate and accurate [7]. The instantaneous growth formula is derived from the equation $Y_t = Y_0 e^{gt}$. To convert a simple annual percentage growth rate P to a continuously compounded instantaneous rate, take the natural logarithm of $(1 + P)$. We are indebted to Philip Sternberg of IBM for helping to sort out the subtleties of these growth calculations.

Using the doubling time allows us to compare the trends in servers to another important parameter popularly reported in this fashion (Moore's law), which in its most precise form states that the number of transistors on a chip doubles roughly every 2 years.[7] The most widely believed incarnation of Moore's law is that performance per microprocessor doubles every 1.5 years, which happens to be true (as documented by Nordhaus 2007), but it is unclear if this popular belief is based on real data or just a misunderstanding of what Moore actually said [8].

9.2.2 Performance

How to measure computing performance has been a source of controversy since the beginning of the computer age, and this chapter will not settle those issues. Each example we developed relies on different performance metrics, but in each case, the performance metric remains consistent over time. It is the time trends that matter for this analysis, not the accuracy of one metric over another.[8]

9.2.3 Energy Use

Energy use of IT equipment has been a major focus of research for more than two decades.[9] The most common error in assessing energy use for computers is to rely on the nameplate power use printed on the computer's power supply, which is generally two to three times larger than typical power use for that device in operation.

We relied mainly on measured data for this analysis, some of which came from SPECpower_ssj2008 [9], available for use since late 2007. As with all benchmarks, it has limitations, but for now, it is the best-available option for associating power use with performance. If we did not have SPEC power runs, we used other measurement methods as available.

9.2.4 Costs

One of the key failings of industry assessments of cost trends in the past is that costs are almost never reported in a form that is consistent with the performance and energy use data. We treated that issue by compiling

---

[7]This "law" has changed in form over the years [8, 10]. At first, Moore [11] referred to "components" not transistors and correctly predicted that the number of components would double every year through at least 1975. In 1975, Moore correctly predicted that the number of transistors on a chip would double every 2 years in the future [12].

[8]Of course, one should always prefer benchmarks that closely approximate real-world workloads when they are available.

[9]For more details, see [13–27], 2003, Piette et al. 1991, Roth et al. 2002.

industry data on equipment prices for configurations of servers for which performance and energy use were reported.

Another issue with costs when they are reported is that they are almost never corrected for inflation. We used the annual implicit deflator for gross domestic product (GDP) from the Bureau of Economic Analysis [28] and the assumption of 2% per year inflation from 2008 to 2009 to adjust all dollar figures to constant 2009 dollars, thus eliminating inflation as a confounding variable in our time trends analysis. When cost data were available by month, we used the monthly GDP deflator data from the Energy Information Administration [29] to correct to July 2009 dollars (assuming 2% average annual inflation expressed as a monthly charge from 2008).

Costs depend on the characteristics of the purchaser, so absolute estimates of the power use per server cost (or other cost-related ratios) are dependent on the particular context in which the servers were purchased. In general (but not always), large purchasers obtain more favorable pricing. In this study, we relied in part on costs produced by online stores for HP, Dell, and IBM. We did not include taxes, shipping costs, software, or service contracts. Where there was a choice, we used costs for small and medium businesses (as distinct from costs for individuals or large corporations).

9.3 RESULTS

Analyzing trends accurately over time requires consistent estimates of performance, server price, and power use per server, measured over a time period sufficient to capture major step changes in chip and server system design. The data also need to be broadly representative of major classes of server applications for lessons derived from them to be generalizable to the industry as a whole. While the examples explored here have limitations, they represent a good first step toward a deeper understanding of trends in server technology.

Figure 9.4 summarizes these quantitative trends in terms of doubling times. The key result is that performance per server costs grew faster than performance per watt in every case analyzed, which explains why watts per server costs continue to increase, as per Equation 9.1b.

9.3.1 Performance

Performance per server generally doubled every 1.5 years or so. The only outlier in the performance data was the Google example from 2001 to 2004, which doubled only every 4 years.

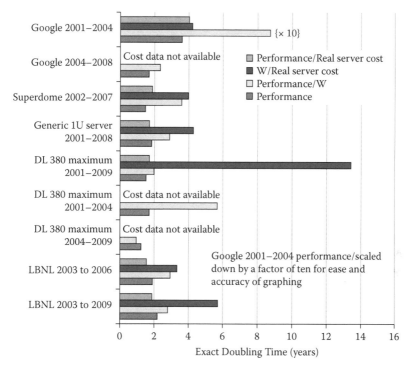

FIGURE 9.4 Summary of trends for servers, expressed as doubling time in years. Longer bars mean slower growth. Doubling time calculated using instantaneous exponential growth rates as described in the text.

9.3.2 Performance/Watt

Performance/watt doubled every 2 to 4 years, again with the exception of Google servers from 2001 to 2004, which showed much slower growth in this parameter than the other examples.

9.3.3 Performance/Server Cost

The Google example from 2001 to 2004 was also the outlier for this parameter. The other examples showed doubling times of about 2 years for performance/server cost. As described, performance per server cost grew more rapidly than performance per watt in all cases.

9.3.4 Watts/Server Cost

Doubling times in watts/server cost were much longer for the DL360 case than for the other examples, but this parameter continued to increase in

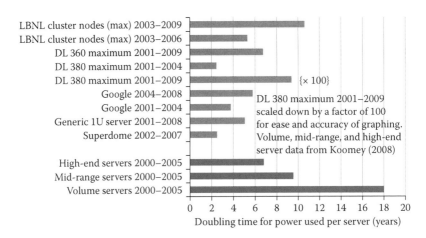

FIGURE 9.5 Doubling times for power used per server (years). For the DL 380 servers, power use per server actually declined from 2004 to 2009, bringing it back to about the 2001 level by 2009 ([5])

all cases. The DL360 case is interesting because the latest model server in this time series was optimized for power efficiency (that was the one available in the SPEC power database). We do not know if the trends derived using these data are representative for the volume server market as a whole.

9.3.5 Watts per Server

All of the previous examples included estimates of doubling times for power use/server. These can be compared to the doubling times contained in the work of Koomey [17] for the server market as a whole (see Figure 9.5). Power use/server appeared to grow faster in the examples presented than in the overall server market. For example, Superdome watts/server doubled in about 2.5 years, which was three times faster than the overall market for high-end servers according to Koomey [17], and the same conclusion held for the other examples compared to the market trend for volume servers.

The market trends included both changes in the power used by consecutive generations of individual server models (like the trends we analyzed in the examples) and shifts in the market share of different server models, which may explain the lower growth rates and longer doubling times in power use/server for the aggregate market data. Further research is needed to validate this inference.

9.4 FUTURE WORK

This analysis was more detailed than any conducted previously, but it is still largely anecdotal in nature. Each example has relevance to some part of the server market, but a more comprehensive approach would be required to accurately understand the aggregate trends. The Superdome server, for example, is only one of many high-end server systems, but high-end machines vary so greatly in their design, construction, and application that more data are sorely needed to better characterize this market. The same lesson holds for the other examples, which fall under the category of "volume" servers.

Future work should therefore include generating more and better server examples, increasing focus on collecting performance data at the applications level, encouraging wider use of energy measurements associated with performance benchmarks, assessing future trends, analyzing underlying technical trends in servers, encouraging technology demonstrations using whole-system redesign, broadening data collection to cover disk drives and network equipment, and assessing the effects of these trends on total data center costs using simple models. For more details, see Koomey [5].

9.5 CONCLUSIONS

As our economy becomes more dependent on computing networks we will need to develop an understanding of the deep underlying trends driving their costs and capabilities. This chapter combined economic data (server costs) with technological information to create consistent comparisons and give insights into the key trends affecting total costs in data center facilities.

Companies that own data centers need to understand the trends affecting true total costs in their facilities. Minimizing costs of computing services requires more than maximizing computing performance per dollar of IT equipment purchased. The direct power used per dollar of IT equipment cost drives the costs of cooling and electricity, which in recent years have come to approach (in annualized terms) the cost of purchasing the IT equipment in many data center facilities.

The data analyzed in this report point to continuing growth in power used per $1,000 of server cost, which will only increase the importance of site infrastructure and electricity costs compared to the cost of IT equipment. This trend places a burden on most companies running enterprise data centers, which have not yet adjusted their design, construction, and operations procedures to reflect this new reality. Incentives within many firms still do not promote minimization of the total costs of delivering computing services, often because total cost is not even analyzed in these companies.[10]

There are technical solutions that can help reduce the cost of computing services, but the problem cannot be solved without changing institutional arrangements and incentives within companies. Split incentives arise when facilities and IT departments have different budgets, or when people using the data center are charged solely per square foot, ignoring power use. Without a simple model of total costs and data assessing underlying trends, it is impossible for companies to understand the full benefits of fixing these institutional problems or moving some of their computing demands to cloud computing providers (which have some inherent advantages in addressing these issues).

In the data compiled here, performance trends for server systems seemed to track the popular interpretation of Moore's law well (it doubled in all but one case every 1.5 to 2 years). In all cases, performance per server cost increased more rapidly than performance per watt, which drives power use per server cost up over time. While there is some evidence that the trend toward increased power use per server cost has moderated in the past few years (because of aggressive efforts by chip and server manufacturers to improve server efficiency), more research will be needed to confirm this conclusion.

ACKNOWLEDGMENTS

This report was produced with grants from Microsoft Corporation and Intel Corporation and independent review comments from experts throughout the industry. All errors and omissions are the responsibility of the authors alone. The affiliations given below were current when the research was completed in 2009, but some funders, colleagues, and reviewers have changed jobs since that time.

---

[10]It is important to distinguish here between the large companies that supply IT services from the companies for which IT is not their core business. Most IT services companies have started down the path of fixing the misplaced incentives and structural problems that impede the minimization of total costs (with differing levels of commitment), but the latter group largely has not. In either case, understanding the trends embodied in the data presented here is critical for improving the design of these facilities.

We would like to thank Mike Manos and Rob Bernard of Microsoft Corporation and Lorie Wigle of Intel Corporation for their financial support of this project. We would also like to thank Mark Aggar (Microsoft) and Scott Shull (Intel) for their technical guidance and Agnies Watson and Lori Blonn for their patient project management assistance.

We would like to thank James Berry of the Midwest ISO for making the HP ProLiant DL360 G1 server available to Intel for testing and Mangesh Tamhankar of Intel Corporation for making Anthony Santos available to work on SPECpower_ssj2008 runs for this analysis and to share his insights on server power trends.

In addition, we are grateful to Gary Jung of the Lawrence Berkeley National Laboratory (LBNL) for making some of his servers available for testing and for assigning an excellent student, Jared Baldridge, to conduct the testing.

SPEC and the names SPECpower_ssj2008 are trademarks of the Standard Performance Evaluation Corporation (SPEC). See www.spec.org.

Finally, we would like to thank the technical reviewers for their insights and comments. The reviewers included (in alphabetical order by company):

AMD: Larry Vertal

AT&T: Michele Blazek

Cisco: Hugh Barrass, Leonid Rabinovich

Dell: Stuart Berke, David Moss, John Pflueger

Ecos Consulting: Chris Calwell

EPA: Andrew Fanara, Katharine Kaplan

Google: Luiz Barroso, Jimmy Clidaris, Bill Weihl

HP: Paul Perez

IBM: Steve Kinney, Joe Prisco, Roger Schmidt

ICF: Arthur Howard, Rebecca Duff, Cody Taylor

IDC: Lloyd Cohen, Vernon Turner

Intel: Henry Wong, Scott Shull

LBNL: Rich Brown, Bruce Nordman, Dale Sartor, Bill Tschudi, Gary Jung

Lumina Systems: Max Henrion

McKinsey: William Forrest

MegaWatt Consulting, Inc.: KC Mares

Microsoft: Mark Aggar, Stephen Berard, Richard Russell

Midwest ISO: James Berry

Massachusetts Institute of Technology: Walt Henry

Rocky Mountain Institute: Amory Lovins

Stanford University: Richard Mount, Phil Reese

Sun Microsystems, Greg Papadopoulos, Mark Monroe, Subodh Bapat

Team HPC: Bret Stouder

Uptime Institute: Ken Brill, John Stanley

Virginia Tech: Kirk W. Cameron, Wu-chun Feng

Yahoo: Christina Page

APPENDIX: SIMPLE COST MODEL FOR DATA CENTERS

The ATC of a data center can be expressed as in Equation 9A.1:

$$ATC = IT + INF_{kw} + INF_{nonkW} + EC + O\&M \qquad (9A.1)$$

where

IT = annualized IT capital;

INF_{kw} = annualized kilowatt-related infrastructure capital;

INF_{nonkW} = annualized non-kilowatt-related infrastructure capital;

EC = annual electricity costs, typically about half for infrastructure and half for direct IT electricity use; and

$O\&M$ = annualized operations and maintenance costs.

To annualize capital costs, we use the capital recovery factor (CRF), defined as

$$CRF = \frac{d(1+d)^L}{((1+d)^L - 1)} \quad (9A.2)$$

where d is the discount rate (7% real), and L is the lifetime of the equipment (3 years for IT equipment and 15 years for infrastructure equipment).

Of course, what we really care about is the cost per delivered computing cycle. For simplicity, let us assume 100% equipment utilization. This means that the maximum number of computations possible for a given data center over the course of a year is the maximum number of operations per second times the number of seconds per year.[11] Dividing both sides of Equation 9A.1 by maximum annual computations, we obtain

$$\frac{ATC}{\text{Annual Computations}} = \frac{IT + INF_{kW} + INF_{non\text{-}kW} + EC + O\&M}{\text{Annual Computations}} \quad (9A.3)$$

Let us assume we will spend $1,000 on IT equipment. That yields annualized costs of $381/year for IT (CRF calculated using a 7% discount rate over 3 years).

Both power-related terms (INF_{kw} and EC) can be expressed as a function of the power use per server cost.

The kilowatt-related infrastructure capital costs can be expressed as

$$INF_{kw} = \$1,000 \times \frac{\$1K}{\$1,000} \times \frac{Watts}{k\ 2009\ \$} \times \frac{kW}{1000\ W} \times \frac{\$24,800}{kW}$$
$$\times CRF\ (7\%\ 15\ yrs) \quad (9A.4)$$

where

watts/$1,000 in 2009 = 79.6 in the base case, based on the data for all IT equipment in a data center found in the work of Koomey et al. (2007), and

$24,800/kW = the capital cost of tier 3 infrastructure in 2009 dollars from Koomey [3], based on Uptime Institute data.

---

[11] Measuring actual utilization and total computational output is complicated. Most data centers produce more than one type of computing, and the costs and value of that computing vary by time of day and sometimes by geography. For the purposes of this simple example, we need not worry about these complexities, but companies wrestling with assessing total costs surely must.

The energy costs can be expressed as

$$EC = \$1,000 \times \frac{\$\,1K}{\$1,000} \times \frac{Watts}{k\,2009\,\$} \times \frac{kW}{1000\,W} \times \frac{8766\,hours}{year}$$

$$\times\,LF \times PUE \times EP \tag{9A.5}$$

where

LF = load factor, defined as average electricity load divided by peak load (typically close to 100% for data centers, although climate variations and other factors can reduce this number to 85–90% in some cases);

PUE = power utilization effectiveness, also known as the site infrastructure energy overhead multiplier; this term characterizes the ratio of total data center electricity use to the IT electricity use, and it is typically about 2.0; and

EP = electricity price, which is around $0.07/kWh for large industrial users in the United States.

The other two terms can be expressed as a fraction of the annualized IT costs, based on the data in the work of Koomey [3].

$$INF_{nonkW} = IT \times 0.19 \tag{9A.6}$$

$$O\&M = IT \times 0.30 \tag{9A.7}$$

These equations were used to make Figure 9.1.

These equations, combined with Equation 9.1 in the main text, can be used to give quantitative insight about the trade-offs among the different cost components of data centers. Let us assume Moore's law drives performance per server cost up by a factor of two over a 2-year period (a doubling time of 2 years). The effect on the power-related components of data center costs depends on what happens to power use per server costs (and, implicitly, to performance per watt).

Table 9A.1 shows several scenarios for data center costs to illustrate the interactions among key parameters. Case 1 corresponds to the costs reported by Koomey [3], which we treat as the base case. Data center facilities vary a lot, but this source is the most well-documented published data on total costs for data centers currently known to us, and it is sufficiently well grounded in current industry practice that relying on it for this schematic example will not lead us too far astray.

TABLE 9A.1   Schematic Cost Calculation for Data Centers (Based on $1,000 of IT Costs)

Case[a]	1 (Base)	2	3	4	5
Performance/real server cost (Base = 1.0)	1.00	2.00	2.00	2.00	2.00
Performance/watt (Base = 1.0)	1.00	2.00	1.00	3.00	0.53
Watts/$1,000 in 2009	79.6	79.6	159.2	53.1	302.8
Arbitrary number of computations	1,000	2,000	2,000	2,000	2,000
Annualized Costs (2009 Dollars)[b]					
IT	$381	$381	$381	$381	$381
Kilowatt-related infrastructure	$217	$217	$433	$144	$824
Electricity costs	$98	$98	$195	$65	$372
Non-kilowatt-related infrastructure	$73	$73	$73	$73	$73
O&M costs	$113	$113	$113	$113	$113
Total	$882	$882	$1,196	$777	$1,763
Index compared to base case	1.000	1.000	1.357	0.881	2.000
Annualized cost per computation	$0.88	$0.44	$0.60	$0.39	$0.88
Index compared to base case	1.000	0.500	0.678	0.441	1.000

(a) Case 1: Base case, circa 2007 for a data center delivering high-performance computing for financial applications (based on model in Koomey et al. 2007).

Case 2: Computations per dollar of IT cost double, as does performance per watt, keeping watts/$1,000 constant.

Case 3: Computations per dollar of IT cost double, and performance per watt does not change, making watts/$1,000 double.

Case 4: Computations per dollar of IT cost double, and performance per watt triples, reducing watts/$1,000 to two-thirds of its base case value.

Case 5: Computations per dollar of IT cost double, and performance per watt declines almost 50% (enough for increased indirect power-related costs to completely offset the IT-related reduction in costs per computation).

(b) IT capital expenditures assumed to remain constant at $1,000. Electricity price = $0.07/kWh. Load factor = 100%. PUE = 2.0. Discount rate = 7% real. Lifetime of IT equipment = 3 years; lifetime of infrastructure equipment = 15 years.

The table examines four other cases. For each of these cases, the ratio of performance to IT costs doubles compared to the base case. We assume that we always spend $1,000 for IT equipment, which implies that total performance will go up by a factor of two (in this example, we rely on arbitrary performance units to simplify the calculations).

For scenario 2, we assume that performance per watt also doubles during this period, implying that watts per $1,000 will remain constant (as per Equations 9.1a and 9.1b). For scenario 3, we assume that performance per watt will remain the same as in the base case, implying that power use per unit of server cost will double. Scenario 4 assumes that performance per watt triples over 2 years, implying that watts per $1,000 of IT cost will reach two-thirds of its value in the base case. Finally, in scenario 5 we calculate the performance per watt (relative to the base case) that would result in the same total cost per computation as in the base case.

REFERENCES

1. Belady, Christian L. 2007, February. In the Data Center, Power and Cooling Costs More than the IT Equipment It Supports. *ElectronicsCooling*, pp. 24–27.
2. Brill, Kenneth G. 2007. *Data Center Energy Efficiency and Productivity*. Santa Fe, NM: Uptime Institute. http://www.uptimeinstitute.org/.
3. Koomey, Jonathan, Kenneth G. Brill, W. Pitt Turner, John R. Stanley, and Bruce Taylor. 2007, September. *A Simple Model for Determining True Total Cost of Ownership for Data Centers*. Santa Fe, NM: Uptime Institute. http://www.uptimeinstitute.org.
4. Patel, Chandrakant D., and Amip J. Shah. 2005, June 9. *Cost Model for Planning, Development and Operation of a Data Center*. Palo Alto, CA: Hewlett Packard Laboratories. HPL-2005-107(R.1).
5. Koomey, Jonathan G., Christian Belady, Michael Patterson, Anthony Santos, and Klaus-Dieter Lange. 2009, August 17. *Assessing Trends over Time in Performance, Costs, and Energy Use for Servers*. Oakland, CA: Analytics Press. http://www.intel.com/pressroom/kits/ecotech.
6. Barroso, Luiz André, and Urs Hölzle. 2007. The Case for Energy-Proportional Computing. *IEEE Computer*, 40(12), 33–37. http://www.barroso.org/.
7. Nordhaus, William D. 2007. Two Centuries of Productivity Growth in Computing. *The Journal of Economic History*, 67(1), 128–159. http://nordhaus.econ.yale.edu/recent_stuff.html.
8. Mollick, Ethan. 2006, July–September. Establishing Moore's Law. *IEEE Annals of the History of Computing*, pp. 62–75.
9. Standard Performance Evaluation Corporation. 2013. SPECpower_ssj2008. http://www.spec.org/power_ssj2008/.

10. Stokes, Jon. 2008, September 27. Classic.Ars: Understanding Moore's Law. http://arstechnica.com/hardware/news/2008/09/moore.ars.

11. Moore, Gordon E. 1965, April 19. Cramming More Components onto Integrated Circuits. *Electronics*, pp. 114–117.

12. Moore, Gordon E. 1975. Progress in Digital Integrated Electronics. *IEEE, IEDM Tech Digest*, pp. 11–13. http://www.ieee.org/.

13. Baer, Walter S., Scott Hassell, and Ben Vollaard. 2002. *Electricity Requirements for a Digital Society*. Santa Monica, CA: RAND. MR-1617-DOE. http://www.rand.org/publications/MR/MR1617/.

14. Blazek, Michele, Huimin Chong, Woonsien Loh, and Jonathan Koomey. 2004. A Data Center Revisited: Assessment of the Energy Impacts of Retrofits and Technology Trends in a High-Density Computing Facility. *The ASCE Journal of Infrastructure Systems*, 10(3), 98–104.

15. Harris, Jeff, J. Roturier, L. K. Norford, and A. Rabl. 1988, November. *Technology Assessment: Electronic Office Equipment*. Berkeley, CA: Lawrence Berkeley Laboratory. LBL-25558.

16. Kawamoto, Kaoru, Jonathan Koomey, Bruce Nordman, Richard E. Brown, Maryann Piette, Michael Ting, and Alan Meier. 2002. Electricity Used by Office Equipment and Network Equipment in the U.S. *Energy—The International Journal* (also LBNL-45917), 27(3), 255–269.

17. Koomey, Jonathan. 2008, September 23. Worldwide Electricity Used in Data Centers. *Environmental Research Letters*, 3(034008). http://stacks.iop.org/1748-9326/3/034008.

18. Koomey, Jonathan, Chris Calwell, Skip Laitner, Jane Thornton, Richard E. Brown, Joe Eto, Carrie Webber, and Cathy Cullicott. 2002. Sorry, Wrong Number: The Use and Misuse of Numerical Facts in Analysis and Media Reporting of Energy Issues. In *Annual Review of Energy and the Environment 2002*. Edited by R. H. Socolow, D. Anderson, and J. Harte. Palo Alto, CA: Annual Reviews (also LBNL-50499), pp. 119–158.

19. Koomey, Jonathan, Huimin Chong, Woonsien Loh, Bruce Nordman, and Michele Blazek. 2004. Network Electricity Use Associated with Wireless Personal Digital Assistants. *The ASCE Journal of Infrastructure Systems* (also LBNL-54105), 10(3), 131–137.

20. Koomey, Jonathan, Mary Ann Piette, Mike Cramer, and Joe Eto. 1996. Efficiency Improvements in U.S. Office Equipment: Expected Policy Impacts and Uncertainties. *Energy Policy*, 24(12), 1101–1110.

21. Lovins, Amory, and H. Heede. 1990, September. *Electricity-Saving Office Equipment*. Snowmass, CO: Competitek/Rocky Mountain Institute.

22. Mitchell-Jackson, Jennifer, Jonathan Koomey, Michele Blazek, and Bruce Nordman. 2002. National and Regional Implications of Internet Data Center Growth. *Resources, Conservation, and Recycling* (also LBNL-50534), 36(3), 175–185.

23. Mitchell-Jackson, Jennifer, Jonathan Koomey, Bruce Nordman, and Michele Blazek. 2003. Data Center Power Requirements: Measurements from Silicon

Valley. *Energy—The International Journal* (also LBNL-48554), 28(8), 837–850.

24. Norford, Les, A. Hatcher, Jeffrey Harris, Jacques Roturier, and O. Yu. 1990. Electricity Use in Information Technologies. In *Annual Review of Energy 1990*. Edited by J. M. Hollander. Palo Alto, CA: Annual Reviews, pp. 423–453.

25. Piette, Maryann, Joe Eto, and Jeff Harris. 1991, September. *Office Equipment Energy Use and Trends*. Berkeley, CA: Lawrence Berkeley Laboratory. LBL-31308.

26. Roth, Kurt, Fred Goldstein, and Jonathan Kleinman. 2002, January. *Energy Consumption by Office and Telecommunications Equipment in Commercial Buildings—Volume I: Energy Consumption Baseline*. Washington, DC: Prepared by Arthur D. Little for the U.S. Department of Energy. A.D. Little Reference no. 72895-00. http://www.eere.energy.gov/.

27. Roth, Kurt, Ratcharit Ponoum, and Fred Goldstein. 2006, March. *U.S. Residential Information Technology Energy Consumption in 2005 and 2010*. Cambridge, MA: Prepared by Tiax for the U.S. Department of Energy, Building Technologies Program. Tiax Reference no. D0295 (Final Report).

28. BEA. 2013. *Implicit price deflators for Gross Domestic Product*, [online]. Bureau of Economic Analysis, U.S. Department of Commerce, 2013 [http://www.bea.gov/iTable/index_nipa.cfm]

29. U.S. DOE. 2013. *Short Term Energy Outlook (Table 9a: US Macroeconomic Indicators and Carbon Dioxide Emissions)*. Washington, DC: Energy Information Administration, U.S. Department of Energy. November 13. [http://www.eia.gov/forecasts/steo/data.cfm?type=tables]

Index

Printed and bound by CPI Group (UK) Ltd, Croydon, CR0 4YY

24/10/2024

01778283-0010